Springer Texts in Education

Springer Texts in Education delivers high-quality instructional content for graduates and advanced graduates in all areas of Education and Educational Research. The textbook series is comprised of self-contained books with a broad and comprehensive coverage that are suitable for class as well as for individual self-study. All texts are authored by established experts in their fields and offer a solid methodological background, accompanied by pedagogical materials to serve students such as practical examples, exercises, case studies etc. Textbooks published in the Springer Texts in Education series are addressed to graduate and advanced graduate students, but also to researchers as important resources for their education, knowledge and teaching. Please contact Nick Melchior at textbooks.education@springer.com or your regular editorial contact person for queries or to submit your book proposal.

More information about this series at https://link.springer.com/bookseries/13812

Melissa Whatley

Introduction to Quantitative Analysis for International Educators

 Springer

Melissa Whatley [ID]
School for International Training
Brattleboro, VT, USA

ISSN 2366-7672   ISSN 2366-7680 (electronic)
Springer Texts in Education
ISBN 978-3-030-93830-7   ISBN 978-3-030-93831-4 (eBook)
https://doi.org/10.1007/978-3-030-93831-4

© The Editor(s) (if applicable) and The Author(s), under exclusive license to Springer Nature Switzerland AG 2022
This work is subject to copyright. All rights are solely and exclusively licensed by the Publisher, whether the whole or part of the material is concerned, specifically the rights of translation, reprinting, reuse of illustrations, recitation, broadcasting, reproduction on microfilms or in any other physical way, and transmission or information storage and retrieval, electronic adaptation, computer software, or by similar or dissimilar methodology now known or hereafter developed.
The use of general descriptive names, registered names, trademarks, service marks, etc. in this publication does not imply, even in the absence of a specific statement, that such names are exempt from the relevant protective laws and regulations and therefore free for general use.
The publisher, the authors and the editors are safe to assume that the advice and information in this book are believed to be true and accurate at the date of publication. Neither the publisher nor the authors or the editors give a warranty, expressed or implied, with respect to the material contained herein or for any errors or omissions that may have been made. The publisher remains neutral with regard to jurisdictional claims in published maps and institutional affiliations.

This Springer imprint is published by the registered company Springer Nature Switzerland AG
The registered company address is: Gewerbestrasse 11, 6330 Cham, Switzerland

To DW and MW, TG, and JM, in order of appearance in my life. None of this would be possible without you.

Preface

The purpose of this book is to introduce international educators to the basics of quantitative analysis and how it can be used to inform and to assess their work in the field. Quantitative analysis is useful to international educators for several reasons. From an applied perspective, quantitative analysis can provide information regarding the outcomes of international education programming for the explicit purpose of improving those programs—often referred to as program assessment. To this end, quantitative analysis can answer important questions such as:

1. Is a program intended to improve access to education abroad for first-generation students impacting participation for the targeted population?
2. Does the program referenced in the previous question unintentionally marginalize other student populations?
3. Do international students on my campus feel welcome?
4. What impact does my institution's international research outreach have on local communities abroad?
5. Does faculty engagement in research with international collaborators help to improve their chances of winning grant funding for their research?
6. Are students who are exposed to internationalized curriculum more likely to gain employment after graduation?

For international educators working within the contexts of individual institutions, such as senior international officers, directors of study abroad and international student services, and faculty members doing the work of curriculum internationalization, these questions are important in understanding how well such efforts address their intended goals and, maybe even more importantly, if they have unintended consequences. Quantitative analysis certainly has much to offer regarding these practice-oriented research questions.

At the same time, international educators must also think more broadly in their critical approaches to analyzing and assessing the field. While single-institution case study research is certainly valuable to the institution itself, how applicable are lessons learned at one institution to a broader audience? What do we know about international education's contribution to economic stratification—in a single country or even at a global level? How do the international flows of students and scholars change over time and, importantly, what does that mean for different regions of the world? How does a nation's political climate produce change in internationalized curricula at higher education institutions? While these broad research questions may not have immediate application to international education in practice, the answers to these questions most certainly have far-reaching implications for international education and a global society generally. Indeed, many of us join the field because we believe that the international exchange of people and ideas can lead to positive outcomes for the world. Quantitative analysis can move us toward answers to these important questions, many of which have implications for the longer-term vitality and impact of our field. As international education professionalizes as a field, it is increasingly essential that scholars, practitioners, and other individuals in international education are able to use quantitative data to assess, evaluate, and critically examine international education activities and practices—both at individual institutions and on broader scales.

Overview of the Book

This book is intended to assist international educators in a variety of professional roles and at all career stages to reach a basic understanding of quantitative research methods and to understand how to use these methods in assessment and research initiatives. Although written for an audience that is based in the USA, the concepts covered in this book are useful for individuals in other regions of the world as well. The chapters that follow guide readers from the most basic of statistical concepts through more complex analytic approaches that are used in quantitative research. When possible, concepts are illustrated using examples from empirical international education research so that readers can see how these approaches are used in practice. For readers interested in learning more about specific statistical approaches beyond what is presented in this book, when applicable, each chapter provides lists of resources that can be used for a deeper dive into the chapter's topic and example studies that apply the statistical approaches presented in the chapter.

Chap. 1 of this book covers basic concepts in quantitative analysis, including data collection, types of variables, and the difference between descriptive and inferential statistics. Chaps. 2 and 3 introduce readers to the basic building blocks of any statistical analysis: measures of central tendency and measures of variability. Measures of central tendency provide researchers with a snapshot of what a variable is like *on average*, while measures of variability provide information about how far away individual data points are from the average. Chap. 4 guides readers through using these pieces of information about variables—average and variability—to test certain assertions (called *hypotheses* in quantitative analysis) about the nature of the relationships among variables in a dataset. For example, do students who take internationalized courses graduate with higher GPAs than those who do not and, importantly, is the difference in average GPA between the two groups greater than random chance? This chapter (Chap.4) introduces a type of hypothesis test called a *t*-test, which is used to explore numerical differences between two groups. Chap. 5 expands on these concepts to introduce two additional statistical tests—one-way ANOVA, which is used to explore numerical differences between three or more groups and the chi-square test of independence, which is used to explore relationships between variables that have categories (i.e., they are not numerical).

Correlation, discussed in Chap. 6, provides an assessment of the association between two variables that are numerical and is a useful stepping stone between simple hypothesis testing and regression analysis. Chap. 7 introduces readers to regression analysis, and specifically Ordinary Least Squares (OLS) regression. Regression allows researchers to incorporate multiple variables into a single analysis. For example, an analysis that only accounts for whether a student takes internationalized courses when explaining variation in academic achievement misses other important information about students—unrelated to internationalized course enrollment—that might also impact academic achievement, such as a student's academic aspirations or first-generation-in-college status. Regression allows for the incorporation of this information into analyses. Chap. 8 expands on Chap. 7 in a number of ways, providing information about additional kinds of variables that can be included in regression analyses and an introduction to another kind of regression, logistic regression. While OLS regression is useful when the outcome variable that a researcher wants to study is numerical, logistic regression is needed when the outcome is binary, meaning that it takes on only two possible values (e.g., whether a student studies abroad or not). Finally, Chap. 9, Introduction to Experimental and Quasi-experimental Design, offers an overview of analytic approaches that have the potential to approach a causal interpretation of the results of statistical analysis. This kind of analysis is useful when we want to make claims that participation in a specific program or experience *causes* a change in students' educational outcomes, for example. The purpose of Chap. 9 is not to provide readers with in-depth information about quasi-experimental research designs—these analytic approaches deserve their own book—but rather to present an overview of these methods so that readers are able to access academic writing about experimental and quasi-experimental research. The final chapter of this book, Chap. 10, discusses how to report quantitative research findings in a variety of contexts, both academic and applied.

A Note on Statistical Software

Throughout this book, considerable space is spent working through specific statistical calculations by hand. The purpose of including this level of detail is so that readers have a full understanding of what goes into these analyses. The inclusion of by-hand calculations is an attempt to demystify statistics that are often presented to readers of empirical work in a non-transparent way. A natural consequence of including by-hand calculations, however, is that numbers are often rounded to simplify the math. For this reason, if you are following along with the calculations (whether in the book text or in the practice problems) and get a slightly different answer from the one provided, it is likely due to differences in rounding.

In reality, researchers almost always make use of statistical software to analyze data. This book does not present analyses using a single, specific software program because of variations in expertise, skill, and preference in the field of international education for certain statistical programs. Many of the calculations presented in this book can be performed in Microsoft Excel or other similar spreadsheet programs. Other common statistical software programs in education research include SPSS and Stata and, in some communities, R. Each of these programs has positive and negative aspects that make them more or less user-friendly, more or less cost-efficient, and more or less transparent in the assumptions that go into a given calculation. To this end, I leave it to readers to explore and select a statistical software program that best suits their needs rather than imposing a particular program.

Brattleboro, USA   Melissa Whatley

Acknowledgements

As with anything worth doing, this book was not written alone, even though my name is the one on the cover. I first thank Robert Sandruck and Brooke Shurer for providing initial thoughts and comments from a practitioner perspective on my ideas for this book. Rosalind Raby encouraged me to explore writing this book early in my career and provided me with initial advice about the book publication process. Santiago Castiello-Gutiérrez offered several well-timed assists helping me find examples of empirical work that applied the statistical approaches that I describe in the book's various chapters and assisted me in constructing the list of journals in Appendix D. I am additionally grateful to Dominique Foster, Timothy Gupton, Jillian Morn, Robert Toutkoushian, and Rachel Worsham, who provided useful comments on specific chapters of this book as it was nearing completion. I also thank the anonymous reviewers who provided critical commentary on both this book's proposal and the full manuscript.

Of course, this book certainly would not have been possible without the outstanding teaching and mentorship in quantitative research methods that I received as a doctoral student in the University of Georgia's Institute of Higher Education. Manuel González Canché deserves all the credit for bringing me over to quantitative methods in the first place. He's the first person who ever told me (over and over) that I was decent at math, and he devoted countless hours to my early formation as a quantitative researcher. Robert Toutkoushian is hands down the best quantitative methods teacher and advisor that a graduate student could ask for. If one day I am half the teacher and advisor that he is, I'll call it a success.

My husband Tim Gupton has supported my efforts writing this book long before I ever typed the first word. Over the course of my still-young career, he's often taken on more than his fair share of household chores and responsibilities all so that I could "go be awesome" (his words, not mine). I honestly don't know how people without such a supportive partner get anything done.

Finally, I thank Jamie Monogan, who planted the initial seed for this book (whether intentional or not), connected me with Springer, and not-so-anonymously reviewed every single word in every single chapter. Without Jamie's well-timed assist, and consistent agreeableness to read my work and answer my questions, this book simply wouldn't exist. At the same time, any description of Jamie's support for my professional endeavors only scratches the surface of what has become my deepest and most meaningful thought partnership. The most important lessons I've learned from Jamie have nothing to do with numbers or equations, but rather push me to explore the vastness and depth of what I've come to understand to be love.

Data Dictionaries for Sample Datasets

The practice problems that accompany many of the chapters in this book make use of sample datasets that readers can find online at (https://sites.google.com/view/melissa-whatley-ph-d/home/introduction-to-quantitative-methods-for-internationaleducators). These datasets derive from publicly available data, and readers are encouraged to visit the Web sites corresponding to each data source to learn more about what they have to offer.

Sample Dataset #1

Sample Dataset #1 contains information about U.S. state flagship institutions (N = 50) in the 2015–16 academic year. This dataset includes information from three sources: the National Center for Education Statistics' Integrated Postsecondary Education Data System (IPEDS) (https://nces.ed.gov/ipeds/) (most data), the Institute of International Education's Open Doors report (https://opendoorsdata.org/) (study abroad participation numbers), and the U.S. Bureau of Economic Analysis (https://www.bea.gov/) (state GDP).

Variable	Description
instnm	Name of the institution
unitid	Unique identifier for each institution that can be used to match the dataset with additional IPEDS data
totalenroll	Total students enrolled for credit during the 12-month reporting period
studyabroad	Number of students who studied abroad
intlstudents	Number of enrolled non-US resident students
locale	1 = rural 2 = town 3 = suburban 4 = urban

Variable	Description
region	1 = New England 2 = Mid East 3 = Great Lakes 4 = Plains 5 = Southeast 6 = Southwest 7 = Rocky Mountains 8 = Far West
hospital	1 = institution has a hospital on campus
instructionexp	Instructional expenditures
GDP	State GDP

Sample Dataset #2

Sample Dataset #2 contains information about U.S. liberal arts colleges (N = 161) in the 2016–17 academic year. This dataset derives from the same two sources as Sample Dataset #1 (IPEDS and Open Doors).

Variable	Description
instnm	Name of the institution
unitid	Unique identifier for each institution that can be used to match the dataset with additional IPEDS data
totalenroll	Total students enrolled for credit during the 12-month reporting period
studyabroad_offered	Whether an institution offered study abroad opportunities
studyabroad	Number of students who studied abroad
SAT_none[a]	1 = An institution did not report an SAT Math 75th percentile score
SAT400499	1 = An institution's SAT Math 75th percentile score fell within the 400–499 range
SAT500599	1 = An institution's SAT Math 75th percentile score fell within the 500–599 range
SAT600699	1 = An institution's SAT Math 75th percentile score fell within the 600–699 range
SAT700plus	1 = An institution's SAT Math 75th percentile score fell within the 700 or higher range
locale	1 = rural 2 = town 3 = suburban 4 = urban
public	1 = Institution classified as public

Data Dictionaries for Sample Datasets

Variable	Description
pctfemale	Percentage of enrolled students that were female
acceptrate	Institution's acceptance rate
pctgrad	Percentage of enrolled students that were graduate students
pctpell	Percentage of enrolled students that received Pell grants

[a]For institutions that reported ACT scores, these scores have been converted to SAT scores using an official College Board correspondence guide (https://www.ets.org/Media/Research/pdf/RR-99-02-Dorans.pdf).

Sample Dataset #3

Sample Dataset #3 contains information about U.S. community colleges (N = 1019) in the 2015–16 academic year. Like Sample Datasets #1 and #2, this dataset derives from IPEDS.

Variable	Description
instnm	Name of the institution
unitid	Unique identifier for each institution that can be used to match the dataset with additional IPEDS data
totalenroll	Total students enrolled for credit during the 12-month reporting period
studyabroad_offered	Whether an institution offered study abroad opportunities
locale	1 = rural 2 = town 3 = suburban 4 = urban
pctover25	Percentage of enrolled students over the age 25
tuition	Amount charged in tuition to in-state students

Equations Summary

Chapter 2

Statistic	Equation	Key
Mean	$\bar{X} = \frac{\sum_{i=1}^{n} X_i}{n}$	X_i: An individual observation i n: Total number of observations

Chapter 3

Statistic	Equation	Key
Variance	$s^2 = \frac{\sum (X_i - \bar{X})^2}{n-1}$	X_i: An individual observation i $\bar{X}$: The mean of X n: Total number of observations
Standard Deviation	$s = \sqrt{\frac{\sum (X_i - \bar{X})^2}{n-1}}$	X_i: An individual observation i $\bar{X}$: The mean of X n: Total number of observations

Chapter 4

Statistic	Equation	Key
Standard Error	$s_{\bar{x}} = \frac{s}{\sqrt{n}}$	s: The sample standard deviation n: Total number of observations
Calculated t for a one-sample t-test	$t = \frac{\bar{X} - \mu}{s_{\bar{x}}}$	$\bar{X}$: The mean of X μ: The hypothesized population mean $s_{\bar{x}}$: The standard error
Confidence intervals	$CI = \bar{X} \pm (t_\alpha)(s_{\bar{x}})$	$\bar{X}$: The mean of X t_α: The critical t-value associated with a particular alpha level $s_{\bar{x}}$: The standard error
Calculated t for a two-samples t-test	$t = \frac{\bar{X}_1 - \bar{X}_2 - 0}{s_{\bar{x}1 - \bar{x}2}}$	$\bar{X}_1$: The mean of X for group 1 $\bar{X}_2$: The mean of X for group 2 $s_{\bar{x}1-\bar{x}2}$: The standard error of the difference between two sample means
The standard error of the difference between two sample means[a]	$s_{\bar{x}1-\bar{x}2} = \sqrt{s_{\bar{x}1}^2 + s_{\bar{x}2}^2}$	$s_{\bar{x}1}$: The standard error for group 1 $s_{\bar{x}2}$: The standard error for group 2
Calculated t for a dependent samples t-test	$t = \frac{\bar{X} - \bar{Y} - 0}{s_{\bar{D}}}$	$\bar{X}$: The mean value at Time 1 $\bar{Y}$: The mean value at Time 2 $s_{\bar{D}}$: The standard error of the differences in means between Time 1 and Time 2
Standard error of the difference in means between Time 1 and Time 2	$s_{\bar{D}} = \frac{s_D}{\sqrt{n}}$	s_D: The standard deviation of the differences in means between Time 1 and Time 2 n: Total number of observations

[a]When sample sizes are roughly equal.

Chapter 5

Statistic	Equation	Key
Sum of squares between groups	$SS_b = \sum_{i=1}^{K} \left[n(\bar{X} - \bar{X}_T)^2 \right]$	$\bar{X}$: The group mean of X $\bar{X}_T$: The grand mean n: Total number of observations
Mean squared between differences	$MS_b = \frac{SS_b}{K-1}$	SS_b: Sum of squares between groups K: The number of groups

Equations Summary

Statistic	Equation	Key
Sum of squares within groups	$SS_w = \sum_{i=1}^{K}\sum_{i=1}^{L}(X_i - \bar{X})^2$	X_i: An individual observation i $\bar{X}$: The group mean of X K: The number of groups L: The total number of observations in each group
Mean squared within differences	$MS_w = \frac{SS_w}{N-K}$	SS_w: Sum of squares within groups K: The number of groups N: The grand total number of observations
Calculated F	$F = \frac{MS_b}{MS_w}$	MS_b: Mean squared between differences MS_w: Mean squared within differences
Expected value in a chi-square test	$E = \frac{(rowsum)*(columnsum)}{N}$	N: The grand total number of observations
Calculated χ^2	$\chi^2 = \sum\left(\frac{(O-E)^2}{E}\right)$	O: Observed value in each cell E: Expected value in each cell

Chapter 6

Statistic	Equation	Key
Covariance	$s_{xy} = \frac{\sum(X-\bar{X})(Y-\bar{Y})}{n-1}$	X: An individual observation of variable x $\bar{X}$: The mean of variable x Y: An individual observation of variable y $\bar{Y}$: The mean of variable y n: Total number of observations
Correlation coefficient	$r_{xy} = \frac{s_{xy}}{(s_x)(s_y)}$	s_{xy}: The covariance between x and y s_x: The standard deviation of x s_y: The standard deviation of y
Calculated t for a correlation coefficient	$t = \frac{r-\rho}{s_r}$	r: Sample correlation coefficient ρ: Hypothesized correlation coefficient (typically set to $= 0$) s_r: Standard error of the sample correlation coefficient
Standard error of the sample correlation coefficient	$s_r = \sqrt{\frac{(1-r^2)}{(n-2)}}$	r: Sample correlation coefficient n: Total number of observations

Chapter 7

Statistic	Equation	Key
Simple linear regression	$\widehat{Y}_i = a + bX_i$ or $Y_i = a + bX_i + e_i$	$\widehat{Y}_i$: Estimated value for the outcome variable Y_i: Outcome variable a: Intercept b: Slope (or coefficient) X_i: Predictor variable e_i: Error term
Slope	$b = \dfrac{(\frac{1}{n-1})\sum(X_i - \bar{X})(Y_i - \bar{Y})}{(\frac{1}{n-1})\sum(X_i - \bar{X})^2}$	X_i: Individual value for the predictor variable $\bar{X}$: Mean value of the predictor variable Y_i: Individual value for the outcome variable $\bar{Y}$: Mean value for the outcome variable n: Total number of observations
Intercept	$a = \bar{Y} - b\bar{X}$	$\bar{Y}$: Mean value for the outcome variable $\bar{X}$: Mean value of the predictor variable b: Slope (or coefficient)
Multiple linear regression	$\widehat{Y}_i = a + b_1 X_{1i} + b_2 X_{2i}$ or $Y_i = a + b_1 X_{1i} + b_2 X_{2i} + e_i$	$\widehat{Y}_i$: Estimated value for the outcome variable Y_i: Outcome variable a: Intercept b_1: Slope (or coefficient) associated with first predictor b_2: Slope (or coefficient) associated with second predictor X_{1i}: First predictor variable X_{2i}: Second predictor variable e_i: Error term
Calculated t for a regression coefficient	$t = \dfrac{b-0}{s_b}$	b: Calculated slope (or coefficient) s_b: Standard error of the regression coefficient

Chapter 8

Statistic	Equation	Key
Logistic regression	$\ln(\frac{P_i}{1-P_i}) = a + bX_i$	$\ln(\frac{P_i}{1-P_i})$: The log odds that the outcome variable will equal 1 a: Intercept b: Slope (or coefficient) X_i: Predictor variable

Contents

1 Introduction to Quantitative Data 1
   1.1   Collecting Quantitative Data 1
   1.2   Surveys 2
   1.3   Populations and Samples 4
   1.4   Discussing Quantitative Data: Types of Variables 6
   1.5   Descriptive and Inferential Statistics 8
   Recommended Reading 9

2 Measures of Central Tendency 11
   2.1   Mode, Median, and Mean 12
        2.1.1   Mode 13
        2.1.2   Median 14
        2.1.3   Mean 15
   2.2   Skewed Distributions 16
   2.3   The Normal Distribution (Part 1) 18
   2.4   Examples from Sample Data and the Literature 19
   2.5   Practice Problems 21
   Recommended Reading 22

3 Measures of Variability 23
   3.1   Range, Variance, and Standard Deviation 24
        3.1.1   Range 24
        3.1.2   Variance 25
        3.1.3   Standard Deviation 26
   3.2   The Normal Distribution (Part 2) 26
   3.3   Examples from Sample Data and the Literature 28
   3.4   Practice Problems 29
   Recommended Reading 30

4 Hypothesis Testing 33
   4.1   Sampling Distribution 36
        4.1.1   Expected Value of the Mean and Standard Error 37
        4.1.2   The t-Distribution 39
   4.2   Hypothesis Testing with One Sample 41

		4.2.1	One-Sample *t*-tests	41
	4.3	Other *t*-Tests		47
		4.3.1	Two-Samples *t*-tests	48
		4.3.2	Dependent Samples *t*-test	50
	4.4	Example from the Literature		52
	4.5	Practice Problems		53
	Recommended Reading			54
5	**One-Way ANOVA and the Chi-Square Test of Independence**			**57**
	5.1	One-Way Analysis of Variance (ANOVA)		58
		5.1.1	Step One: Between-Group Variation	59
		5.1.2	Step Two: Within-Group Variation	60
		5.1.3	Step Three: The F-distribution	62
		5.1.4	Example from the Literature	64
	5.2	Chi-Square Test of Independence		65
		5.2.1	Calculating Expected Frequencies	67
		5.2.2	Comparing Observed and Expected Frequencies	68
		5.2.3	The Chi-Square (χ^2) Distribution	68
		5.2.4	Example from the Literature	71
	5.3	Looking Forward		72
	5.4	Practice Problems		72
	Recommended Reading			74
6	**Correlation**			**75**
	6.1	Correlation: Main Ideas		75
	6.2	Calculating Pearson's Correlation Coefficient		79
	6.3	Additional Correlation Calculations		81
		6.3.1	The Coefficient of Determination	81
		6.3.2	Significance Testing	82
	6.4	Examples from Sample Data and the Literature		83
	6.5	Correlation and Causation		87
	6.6	Practice Problems		87
	Recommended Reading			89
7	**Ordinary Least Squares Regression**			**91**
	7.1	Simple Linear Regression		92
		7.1.1	Functional Form	94
		7.1.2	Ordinary Least Squares	95
		7.1.3	Calculating a Simple Linear Regression	98
		7.1.4	Regression Residuals and Error	100
	7.2	Multiple Linear Regression		100
		7.2.1	Partialling Out	102
		7.2.2	Full Function for Multiple Regression	104
		7.2.3	Hypothesis Testing for Regression Coefficients	104
		7.2.4	Model Fit	106
	7.3	Example from the Literature		107

	7.4	Another Note on Correlation and Causation	109
	7.5	Practice Problems	110
		Recommended Reading	111
8	**Additional Regression Topics**		**113**
	8.1	Categorical Predictors	114
	8.2	Interactions Between Variables	118
	8.3	Functional Form	123
		8.3.1 Quadratic Functional Form	125
	8.4	Binary Outcome Variables	128
		8.4.1 Logistic Regression	130
		8.4.2 Odds Ratio	133
		8.4.3 Average Marginal Effects	134
		8.4.4 Pseudo-R^2	134
	8.5	Example from the Literature	135
	8.6	Looking Forward	137
	8.7	Practice Problems	137
		Recommended Reading	140
9	**Introduction to Experimental and Quasi-Experimental Design**		**141**
	9.1	Experimental Design: Randomized Control Trials	142
		9.1.1 Treatment and Control Groups	142
		9.1.2 Randomization and Selection Bias	143
		9.1.3 Outcomes	145
		9.1.4 Threats to Validity	146
		9.1.5 Example from the Literature	147
	9.2	Nonexperimental Contexts	148
		9.2.1 Propensity Score Matching	149
		9.2.2 Example from the Literature	152
	9.3	Quasi-Experimental Design	154
		9.3.1 Regression Discontinuity	155
		9.3.2 Difference-In-Differences	158
	9.4	Future Quantitative Research in International Education	165
		Recommended Reading	165
10	**Writing About Quantitative Research**		**167**
	10.1	Scholarly Writing	167
		10.1.1 Introduction	168
		10.1.2 Theoretical Framework and Literature Review	170
		10.1.3 Method	171
		10.1.4 Results	174
		10.1.5 Limitations	176
		10.1.6 Discussion and Conclusion	177
	10.2	Writing for External Audiences	177
Practice Problems Answer Key			**179**

A Critical Values of the t Distribution 207
B Critical Values of the F Distribution at $\alpha = 0.05$ 209
C Critical Values of the Chi-Square Distribution 219
D Current International Education (and Related) Journals 221
 Glossary 223
Bibliography 229

Introduction to Quantitative Data

The first step in conducting quantitative analysis is to acquire data that lend themselves to this analytic approach, which can be broadly defined as analysis that involves numerical data. This first chapter introduces key issues in quantitative data collection, including survey design and sampling strategy, as well as key terminology that we will use to talk about quantitative data throughout this book. The concepts discussed in the following sections form the foundation for the analytic approaches outlined in subsequent chapters. This first chapter ends with a brief introduction to two main approaches to analyzing quantitative data, descriptive and inferential statistics.

1.1 Collecting Quantitative Data

Quantitative data, that is, data involving numbers, can be collected in several ways and from a variety of sources. Many quantitative data sources do not require researchers to collect data themselves. For example, at institutions of higher education in the United States, individual offices and units on campus are usually responsible for collecting data about the institution. These campus units have names such as 'Institutional Research' or 'Institutional Effectiveness.' Partnering with individuals in these units can be a first step in acquiring international education data at a single institution. More broadly, governments and international education organizations also regularly collect data that are useful for international educators. In the United States, these sources might include the U.S. State Department, the Institute of International Education, and the National Center for Education Statistics. Many other nations also make available similar data. Globally, the World Bank collects data on nation-level finances and UNESCO shares a large quantity of education-related data on its website. These data sources are but a few examples of the numerous quantitative data resources currently available to international education researchers.

1.2 Surveys

However, existing data sources are sometimes not sufficient to address a researcher's specific questions. In this case, the researcher must collect data themselves. Survey data is a common means of collecting data in international education research, and potential survey participants include students, international education professionals, faculty, or community members that come into contact with international education activities in some way. While issues in collecting data through researcher-written surveys are deserving of their own book, a few key recommendations regarding survey data collection are important in this book because they impact the quality and usefulness of the data that a researcher collects[1]:

1. *Begin with an objective (or two) in mind.* It may help to write these objectives down both to refine your goals and to remind yourself why you are collecting survey data. The questions that you include in your survey should tie back to your objective—if they do not, then the data you collect is not likely to help you address its intended purpose.
2. *Consider sharing your objective with your survey participants.* People are often more likely to fill out a survey when they know why they are filling it out. As we will soon see, in much of quantitative research, a researcher is better off when datasets are large.
3. *Keep your survey as short as possible.* It is important to keep your objectives in mind as you write your survey. Ask questions that are relevant while also respecting the time of your participants. Remember that your participants are probably overwhelmed with survey requests. Again, your goal is to get as many individuals to respond to your survey as possible.
4. *Use questions that have already been used by others.* When using a survey to measure complex constructs, such as *intercultural understanding, satisfaction with campus climate*, or *global perspectives*, researchers are encouraged to consider measurement instruments that already exist in the field before authoring their own instruments. For example, a researcher interested in exploring students' development of global perspective-taking might consider using the Global Perspectives Inventory (often referred to as the GPI; Braskamp et al., 2009). Such data collection instruments have often already been widely tested in the literature, thus providing an empirical basis for their quality.
5. *Write for your audience.* Keep your audience in mind as you write your survey. Some audiences will have a greater tolerance for longer surveys than others. Similarly, different kinds of questions will appeal to different audiences.

---

[1] I am grateful to the NAFSA International Students and Scholars Services Knowledge Community, and especially Jana Jaffa and Alena Palevitz, for their feedback on this list of survey design recommendations. An earlier version of this list was used in a NAFSA Collegial Conversation (Survey Design 101: Collecting Effective Survey Data in International Education) that took place on December 15, 2020.

1.2 Surveys

6. *Consider different types of questions.* While multiple choice and short answer questions may be the first to come to mind, consider other question types, such as rank order or a matrix table that might be more appropriate for your questions (See Table 1.1 for examples). Generally speaking, if you can avoid participants writing answers themselves, the data you collect will be easier to analyze, at least from a quantitative perspective.
7. *Collect demographic information.* Regardless of your reason for collecting survey data, always collect information about your participants' demographics. You may think it unnecessary, but chances are, you will need this information at some point during the research process. When collecting demographic data, use inclusive language. When your survey participants feel included and that their voices are being heard, they provide you with higher-quality responses.
8. *Use a survey data collection tool.* Tools like Qualtrics, Survey Monkey, and even Google Forms can greatly facilitate survey data collection. These tools

Table 1.1 Types of survey questions and examples

Question type	Example
Multiple choice	Please select your gender identity: Man Woman Trans-man Trans-woman Gender queer Other: _____ Prefer not to respond
Short Answer	In a sentence or two, please describe the role that research plays in your current professional position.
Rank order	Please rank the following reasons in order of importance (4 = most important and 1 = least important) for why you chose your study abroad program: ___ Location ___ Classes offered ___ Faculty member encouragement ___ Friend encouragement
Matrix table	Please review the statements below. For each statement, use the scale to indicate the extent to which you agree or disagree |	Strongly agree	Agree	Neutral	Disagree	Strongly disagree
---	---	---	---	---	---
I like listening to people from other places, cultures, or countries.					
I try to respect people from other places, cultures, and countries.						

make taking your survey easy for your participants and also facilitate data analysis for you. Consider platforms that are user-friendly for participants in different countries (if applicable) and be mindful about the timeframe during which you collect data (e.g., different holidays or academic calendars around the world).
9. *Pilot your survey.* Ask a few people, preferably those similar to your actual survey takers, to review your survey for clarity before you send it out. For example, if your survey is for students, ask a few students for feedback on your survey. Clearly, if survey participants do not understand your survey questions, they cannot provide you with reliable responses.
10. *Share back.* Once you have collected and analyzed your survey data, share your results with the people who participated and/or others who helped you collect the data. This builds goodwill with others for the next time you need to collect data.

Regardless of how you end up acquiring data, whether from an already-existing data source or a survey, data collection almost always involves the use of a sample from a particular population of interest (e.g., international students, students who have studied abroad, etc.). The next section discusses various strategies used to sample data from a population.

1.3 Populations and Samples

A **population** represents the complete group of units (individuals, institutions, countries, etc.) that is the focus of a particular study while a **sample** is the subset of these units that is used in a specific research study. In an ideal situation, a researcher might want to have population-level data, meaning that each unit that exists in the real world is also represented in the researcher's dataset. However, for numerous practical reasons, this ideal situation is rarely possible. That is, it is hardly ever practical to collect data from every single member of a certain population and, even if it were, it is rarely likely that every single unit is willing to participate in a research project. To that end, researchers employ numerous **sampling strategies** to collect data.

Sampling strategies. A sampling strategy is the process that a researcher uses for collecting data from a subset of the population of interest in a given study. Three primary sampling strategies are employed in quantitative research, presented here from the most to the least rigorous: random sampling, representative sampling, and convenience sampling.

Random sampling. Sampling in a way that each member of a study's population of interest has equal likelihood of being selected to participate in the study is called **random sampling**. An example of random sampling is when students' e-mail addresses are selected at random to receive a survey completion request. For example, I might distribute a campus climate survey to a randomly selected sample of 100 international students enrolled at my institution. The *random* part of this

1.3 Populations and Samples

sampling strategy is important because it means that the subsequent data that are collected do not favor one specific subset of the population of interest. In other words, the findings that result from a study that uses random sampling are not *biased* in favor of a particular group of participants. Each individual international student has an equal likelihood of being represented in the study's results.

Importantly, this unbiased character of a random sample is only as good as the response rate on the researcher's data collection instrument, such as a survey. That is, a researcher might recruit a random sample of participants for a study, but individuals' responses might be biased based on who responds to the invitation to participate. This particular case of bias, called *response bias*, is important to consider, especially in studies where the researcher collects data themselves. For example, in a study of survey response rates among community college students, one study found that women were more likely to complete surveys compared to students representing other gender groups (Sax et al., 2008). Such overrepresentation of one group and underrepresentation of another has the potential to bias the results that researchers might derive from a dataset. To this end, a researcher should report a data collection instrument's response rate and provide demographic information about both the population of interest (if possible) and the study's sample participants to show that they are, in fact, similar.

Representative sampling. One issue with random sampling is that the sample selected from a population at random might unintentionally exclude a certain subset of the population that the researcher is especially interested in. For example, it is possible that a true random sample from a population of international students studying in the U.S. would contain very few—if any—students representing countries with low representation in the international student population itself. Researchers can address this issue through **representative sampling**, wherein the researcher purposefully selects certain individuals or units from the population so that the larger population is proportionally represented. For example, if I am especially interested in documenting the experiences of students from Morocco who are enrolled at my institution, but I know that Moroccan students represent only 5% of enrolled international students, I might purposefully reach out to randomly selected students from Morocco and ask them to participate in my survey, with the goal of ensuring that at least 5% of my sample is Moroccan.

While representative sampling can ensure that certain key groups of the population are represented in the researcher's dataset, it can also be very expensive and time-consuming. Moreover, there is no guarantee that individuals from groups that comprise a smaller proportion of the population will be willing to participate in a particular research project.

Convenience sampling. A final means of collecting data is **convenience sampling**. Unfortunately, this sampling strategy is often the most frequent in international education research—for example, when a researcher collects data at their own institution or from students participating in a program that they themselves organize or lead. Convenience sampling is what happens when a researcher collects data from participants that are in close proximity and willing to participate. While this sampling strategy is certainly the easiest for the researcher in terms of both time and money, it also has the potential to produce the most bias. That is, individuals within close proximity to a researcher might represent only a subset of the population a researcher wishes to study. An example of this situation might be if a researcher wants to study a particular international education issue as it applies to all educational institutions, but only collects data from their own institution. Possibly a more worrisome issue with convenience sampling is that individuals' willingness to participate in a study could be the result of potential coercion—students might feel especially pressured to participate in a study if the person collecting the data is their instructor (or another individual in some sort of authority position). Moreover, it is possible that individuals respond to surveys in a different way when they have a personal or working relationship with the researcher collecting the survey data. For example, recent research by Johnson (2018) suggests that students who have studied abroad emphasize certain aspects of their experiences abroad and fail to mention others when recounting these experiences to an instructor on the study abroad program. In sum, while convenience sampling is a tempting means of data collection because it is easy, it is only acceptable when the sample represents the population that a researcher intends to study and when potential participants do not feel coerced to participate or alter their participation in a certain way because of their relationship with the researcher.

1.4 Discussing Quantitative Data: Types of Variables

Once you have collected data, you will find that you have a collection of information about each of the participants in your study. In quantitative analysis, these pieces of information are called variables. A **variable** represents a certain property or characteristic that corresponds to individual units (or observations) in a dataset. In spreadsheet format, units are often stored in rows while variables are stored in columns. For example, variables in a dataset of countries (rows) might be the continent where each country is located or each country's GDP (stored in columns). A dataset at the student level might contain demographic information about each individual student and information about their participation in a variety of international education opportunities. The following paragraphs define several different types of variables, each of which will be important as we consider different approaches to analyzing quantitative data. A summary of variable types can be found in Table 1.2.

Variables come in two basic formats—categorical and continuous. A **categorical variable** is one that has categories—such as race/ethnicity, gender, socioeconomic status, or other demographic information that might apply to students (some researchers refer to categorical variables as *discrete variables*). Within the broader division of categorical variables is a special kind of variable, a **dichotomous variable**. Dichotomous variables are categorical variables that only have two values—such as whether a student studied abroad or if a country's official language is English (both of these variables can be answered either *yes* or *no*). On the other hand, a **continuous variable** is numerical and indicates a specific

1.4 Discussing Quantitative Data: Types of Variables

Table 1.2 Types of variables: definitions and examples

Type	Definition	Example
Categorical	A variable that has categories	Race/ethnicity
Dichotomous	A categorical variable that has only two categories	Did a student participate in study abroad?
Continuous	A numerical variable that indicates a specific quantity	A country's GDP
Nominal	Numbers represent categories of a variable and are not meaningful	A student's home region: 1 = Africa 2 = Asia 3 = Australia 4 = Europe 5 = North America 6 = South America
Ordinal	Numbers are assigned to categories of a variable and can be used for ordering those categories in a way that is meaningful. However, the distance between categories of a variable is subjective.	A student's level of satisfaction with an international program: 1 = Not satisfied 2 = Somewhat satisfied 3 = Satisfied 4 = Very satisfied
Interval/Ratio	Numbers can be used to order observations in a way that is meaningful, and the difference between values along the scale is uniform	Number of kilometers from a student's home to their host institution

quantity. A country's GDP, an individual's income, or the amount an educational institution charges in tuition are all examples of continuous variables.

While categorical and continuous variables are one way in which quantitative researchers discuss information that is contained in datasets, other ways of describing variables are also useful. **Nominal variables** are a kind of categorical variable wherein information is often coded numerically, but numbers are representative of categories and are otherwise not meaningful. A dataset containing information about international students, for example, might contain information about each student's home region, with each region assigned a number. For example, the number 1 could correspond to Africa, while 2 and 3 correspond to Asia and Australia, respectively. Importantly, these numbers are arbitrary. Other than the convention of putting this sort of information in alphabetical order, there is no inherent reason why Africa is number 1 and Asia is number 2. These numbers are not intended to rank world regions or provide any other sort of information about them. Numbers simply help keep the dataset organized and help perform subsequent analyses. It is important for researchers to be wary that this kind of variable might exist in the dataset they are working with so that they do not inadvertently represent these variables as meaningful numbers (e.g., finding the average of a nominal variable is possible, but meaningless) or apply analytic approaches that require a continuous rather than a categorical variable.

Ordinal variables represent another variable type. Similar to nominal variables, these variables represent categories. However, unlike nominal variables, the numerical categories of ordinal variables are meaningful. Likert-scale survey response options (that is, items where a survey respondent is asked how much they agree with a given statement) are good examples of ordinal variables. For example, when we ask a student to rank their level of satisfaction with an international program, we are asking them to select the category that best describes their level of satisfaction. On our scale, 1 might represent a total lack of satisfaction (*not satisfied*), 2 might represent *somewhat satisfied*, while 3 and 4 represent *satisfied* and *very satisfied*, respectively. These categories, then, when put in order (1, 2, 3, 4) represent increasing levels of satisfaction. In other words, the ordering of numerical categories is meaningful. However, the difference between a 1 and a 2 on this scale might be larger than the difference between a 3 and a 4. These categories might even be interpreted differently by different students. For this reason, researchers need to consider how they use and interpret these variables very carefully when they are included in statistical analyses, taking into account their categorical nature.

A final type of variable is an **interval or ratio variable**. Like ordinal variables, these variables can be ordered in a way that is meaningful. However, in contrast to ordinal variables, the distance between one number and the next is the same along the whole scale. An example of an interval/ratio variable is the number of kilometers from a student's home to their host institution abroad. A student who attends an institution 346 km from their home is further from home than one who attends an institution 254 km from home. Moreover, the amount of distance between each kilometer is the same—that is, the difference between 34 and 35 km is the same as the difference between 163 and 164 km. While some researchers differentiate between interval and ratio variables in that ratio variables have a *meaningful* zero value whereas interval variables do not, this distinction is not important for the purposes of this book.[2]

1.5 Descriptive and Inferential Statistics

When collecting data and organizing variables, researchers need to consider what kind of quantitative analysis—that is, what kind of statistics—will be usefully applied to the data. The distinction between a sample and a population is important when considering whether descriptive or inferential statistics are needed to answer a particular research question. **Descriptive statistics** are useful in that they describe a particular sample. Using descriptive statistics, a researcher provides the

---

[2] For the reader interested in the interval vs. ratio difference, consider that "kilometers from home" is a ratio variable because (in principle) a student could attend school 0 km from home. This actually makes multiplication and division also valid in that a student who is 300 km from home is *twice* as far as one who is 150 *km* from home. By contrast, temperature on a Celsius scale (much less Fahrenheit), an interval variable, does *not* have a meaningful zero–even setting 0 at the freeze point is arbitrary. Thus, we could *not* say 50° is twice as hot as 25°.

audience with a general summary of the variables that they use in their analyses, information that is essential for understanding the study that a researcher conducted. Descriptive statistics are also useful for finding errors in a dataset, such as when a variable's average is too high or too low to be valid or if the maximum value of a variable that should only have values between 1 and 4 is 42. However, descriptive statistics are limited to describing a particular sample, and researchers are often interested in conducting analyses that generalize to a larger population. This is where **inferential statistics** come in. In inferential statistics, researchers make educated guesses, based on previous literature or theory, about a population and then use sample data to explore the likelihood that these guesses represent reality. In statistics vocabulary, these educated guesses are called hypotheses. For a researcher to test hypotheses and make inferences about a population based on a given sample, a sample must be representative of that population, and we must be willing to make certain assumptions about that population (more about these assumptions will be introduced throughout the beginning chapters of this book).

Recommended Reading

A Deeper Dive

Agresti, A., & Finlay, B. (2009a). Introduction. In *Statistical methods for the social sciences* (pp. 1–10). Pearson.

Agresti, A., & Finlay, B. (2009b). Sampling and measurement. In *Statistical methods for the social sciences* (pp. 11–30). Pearson.

Braskamp, L. A., Braskamp, D. C., & Merrill, K.C. (2009). Assessing progress in global learning and development in students with education abroad experiences. *Frontiers: The Interdisciplinary Journal of Study Abroad, 18,* 101–118.

de Leeuw, E.D., Hox, J.J., & Dillman, D.A. (Eds). (2008). *International handbook of survey methodology*. European Association of Methodology.

Johnson, K. M. (2018). Deliberate (Mis)representations: A case study of teacher influence on student authenticity and voice in study abroad assessment. *International Journal of Student Voice, 3,* 4.

Sax, L. J., Gilmartin, S. K., Lee, J. J., & Hagedorn, L. S. (2008). Using web surveys to reach community college students: An analysis of response rates and response bias. *Community College Journal of Research and Practice, 32*(9), 712–729.

Urdan, T. C. (2017). Introduction to social science research principles and terminology. In *Statistics in plain English* (pp. 1–12). Routledge.

Measures of Central Tendency

Once you have collected data, you may find yourself in a place where you are wondering what to do next. Of course, you probably collected data with a specific purpose in mind, such as to explore whether taking globalized courses improves students' learning outcomes or to examine the environmental impact of student and scholar mobility on receiving countries. Regardless of your research question or the complexity of the analysis that you will eventually need to answer it, describing key variables in your study numerically, and sometimes visualizing their distributions, is a necessary first step in any quantitative research project. In statistics, when we talk about the **distribution** of a variable, we are referring to the frequency with which each value of that variable appears in the dataset. Figure 2.1 provides an example of a frequency distribution in what is often referred to as a **histogram**. This figure was produced using one of the sample datasets that is used throughout this book (Sample Dataset #1). More information about these sample datasets can be found at the beginning of the book in a section entitled Data Dictionaries for Sample Datasets.

The histogram in Fig. 2.1 displays international student enrollment at U.S. flagship institutions ($N = 50$) from the 2015–16 academic year. In a histogram like this one, the vertical axis refers to the frequency with which observations representing particular values of a variable occur in a dataset. In our example, this axis represents counts of flagship institutions. The horizontal axis divides data into categories, often referred to as 'bins', that are of equal sizes. In Fig. 2.1, bins are in intervals of 2000, meaning that the first category includes flagship institutions that enroll between 0 and 1999 international students, the second category includes flagship institutions that enroll between 2000 and 3999 international students, and so on. In Fig. 2.1, we see that over 20 institutions fall into the 0–1999 category and somewhere between 10 and 15 fall into the 2000–3999 category.

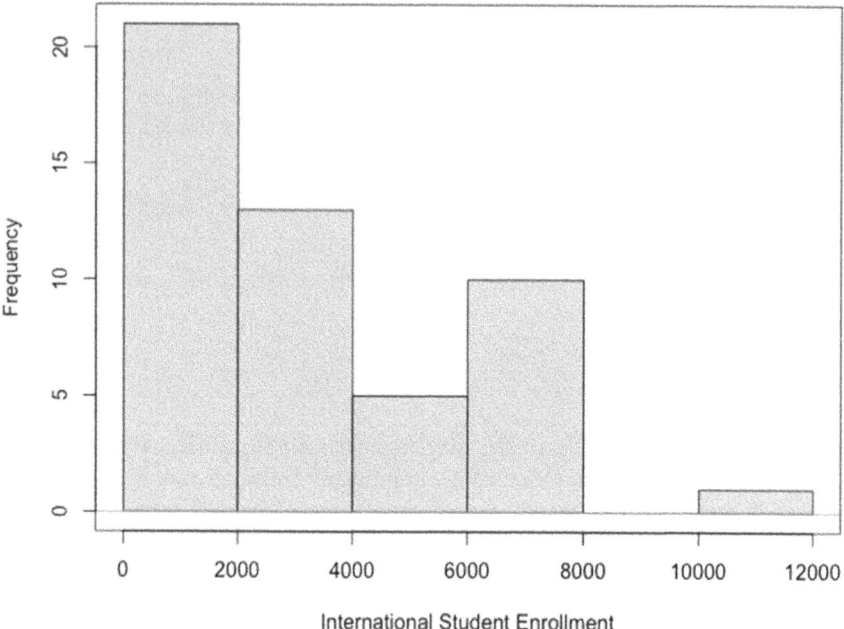

Fig. 2.1 Histogram summarizing international student enrollment at U.S. flagship institutions from the 2015–16 academic year. (*Data source:* Sample dataset #1 - US National Center for Education Statistics, Integrated Postsecondary Education Data System)

Examining key variables in your study, whether numerically or visually, helps you get to know your data, an important part of any research project, and helps to identify potential errors in your dataset. For example, if I know that a variable was measured on a scale between 1 and 5, I know that values higher than 5 or lower than 1 are likely errors or evidence of missing data in my dataset. This chapter (Measures of Central Tendency) and the next (Chap. 3: Measures of Variability) provide you with the tools you need to examine and describe the variables in a dataset.

2.1   Mode, Median, and Mean

Measures of central tendency provide a quick summary of a variable, answering the question: *What is the typical unit (individual, institution, country, region, etc.) like?* There are actually three measures that can answer this question: the mode, the median, and the mean, depending on how you think about the definition of *typical*.

2.1.1 Mode

The most frequent number in a variable's distribution is the mode. For example, suppose we are interested in the number of globalized courses a student takes during their first year of enrollment at an institution. We might randomly select ten students and request their transcripts, resulting in a dataset that looks like Table 2.1 . In this example, Students 1 and 2 did not take any globalized courses during the first year of their studies, Student 3 took four globalized courses, and so on. The mode is simply the number that appears most frequently in the 'Number of Globalized Courses' column. In this case, Students 1, 2, and 5 took no globalized courses in their first year of enrollment, making 0 the mode for this variable. This information might be important to know—whether for research or decision-making purposes. The majority of students appear to not access globalized courses at all. However, we also might argue that 0 is not a particularly good summary of this distribution—after all, some students took as many as 7 or 8 globalized courses in their first year. For some distributions, the mode will be more informative than for others. In fact, for categorical variables (e.g., a student's home region), the mode is really the only meaningful measure of central tendency.

Since there are only ten students in our example, finding the mode of this distribution is relatively easy—we really can just eyeball it. However, with larger datasets, you'll want to make use of what's called a **frequency table**, which is very similar to a histogram, but in table form. For illustrative purposes, I've reorganized the data in Table 2.1 into a frequency table, found in Table 2.2. Most statistical software programs can make a frequency table for you, so you will likely never have to make one yourself by hand. A typical frequency table has four columns. The first contains categories of your variable of interest – in this example, we use the seven categories that correspond to the number of globalized courses that a student took ('0 courses', '1 course', etc.). However, with continuous variables, you might want to create your own categories. For example, if the

Table 2.1 Globalized courses taken by students in their first year

Student	Number of globalized courses
1	0
2	0
3	4
4	8
5	0
6	5
7	7
8	1
9	1
10	3

Note Data in this table are invented for illustrative purposes

Table 2.2 Frequency table for globalized courses taken by students in their first year

Number of globalized courses	Frequency	Percent (%)	Cumulative percent (%)
0 courses	3	30	30
1 course	2	20	50
3 courses	1	10	60
4 courses	1	10	70
5 courses	1	10	80
7 courses	1	10	90
8 courses	1	10	100
Total	10	100	100

Note Data in this table are invented for illustrative purposes

variable you are interested in is number of enrolled international students, it might make sense for these categories to be in thousands, similar to the histogram in Fig. 2.1 (e.g., '0–1,000 international students', '1001–2000 international students', '2001–3000 international students', etc.). The second column of a frequency table, called 'Frequency', contains the number of observations (in our case, students) that fall into each category. The third column lists the percentage of total observations in that category. In this example, since three out of ten students took no globalized courses, Table 2.2 tells us that 30% of students fall into the '0 courses' category. Moreover, we know that '0 courses' is the mode of our distribution since this is the highest percentage in the third column of Table 2.2. Finally, the last column of a frequency table contains the cumulative percent, meaning that it adds up the percentages consecutively in each row. In the second row of this table (the '1 course' category) the cumulative percent is 50%, meaning that half of the students in our dataset took one or fewer globalized courses. Notice that this column is only useful if categories can be organized in a meaningful order (e.g., from fewer to more internationalized courses in this example).

2.1.2 Median

We just mentioned that while the mode is often a useful measure of central tendency, it is not always the best indicator of what the typical unit in a dataset is like. Both the median and the mean are other options to help describe variables. A variable's **median** is the number in the middle of its distribution once you have placed the numbers in order from lowest to highest. Calculating the median for a variable with an odd number of cases is straightforward. For example, if a variable has five cases, say 4, 2, 8, 9, and 1, all we have to do to find the median is to sort the cases in numerical order (1, 2, 4, 8, 9) and find the number in the middle (in this case, 4). This number is the median. If the number of cases is even, as in the example data in Table 2.1 (there are 10 students), then finding the median takes a

2.1 Mode, Median, and Mean

bit more work. After sorting the cases numerically:

$$0, 0, 0, 1, 1, 3, 4, 5, 7, 8$$

We find the two values that fall in the middle of the distribution. In this case, those two values are 1 and 3. Then, we add them together ($1 + 3 = 4$) and divide by 2, taking the average, to get the median value (arriving at 2, in this case).

The median is an especially useful measure when a variable's distribution has a few particularly high or low values, called **outliers**. Because these outliers are not especially representative of the *average* unit, we might not want them to impact the measure of central tendency that we report. For example, imagine that instead of taking eight globalized courses in their first year, Student 4 in Table 2.1 had instead somehow managed to take 15 globalized courses (perhaps through some sort of unique immersion abroad program). This student would not be very typical of most of the students in our dataset. If we were to calculate the median for this distribution:

$$0, 0, 0, 1, 1, 3, 4, 5, 7, 15$$

We would still arrive at 2 ($1 + 3 = 4$ divided by 2). This median value does not change even if one student took an excessive number of globalized courses. It is therefore robust against extreme values.

2.1.3 Mean

The mean of a distribution tends to be the most frequently reported measure of central tendency because it is the most useful in many situations. The **mean** is simply the arithmetic average of a distribution. In the case of our globalized courses example (Table 2.1), we simply add up all the values:

$$0 + 0 + 4 + 8 + 0 + 5 + 7 + 1 + 1 + 3 = 29$$

and then divide by the number of units (in this case, students) in the dataset ($29 \div 10$) to arrive at the mean (2.9 in this case). In our example, on average, students took 2.9 globalized courses. This mean is likely a good summary of what the typical student in our dataset is like as far as globalized course-taking goes, and this is the case for many distributions. The formal equation for finding the mean looks like this[1]:

$$\overline{X} = \frac{\sum_{i=1}^{n} X_i}{n}$$

Here, the mean (represented $\overline{X}$) is equal to the sum of all individual observations (X_i) divided by the total number of observations (n). The symbol $\sum_{i=1}^{n}$ (sometimes called a "summation sign") is used as shorthand to represent that we are summing all observations, starting with the first one ($i = 1$) up to the last observation in the dataset (n). If we only wanted to sum the first five observations in the dataset, we could reflect that decision by changing this symbol to look like this: $\sum_{i=1}^{5}$.

The mean is often the best way to summarize a distribution, providing a useful answer to the question, *What is the typical unit like?* However, unlike the median, it is subject to the influence of outliers. If we return to our example where Student 4 took 15 globalized courses instead of eight, the sample mean is almost an entire unit higher (3.6) compared to the mean we just calculated (2.9). Reporting that students took a mean of 3.6 globalized courses in their first year in this case could be misleading, as it gives the impression that the average student took almost an entire term's worth of globalized courses (assuming that the norm at this institution is that full-time students take 4–5 courses per term). If we view this distribution of number of globalized courses (modified to include the student who took 15 globalized courses), displayed in Fig. 2.2, we see that 3.6 may not be a particularly accurate depiction of the number of globalized courses that the *typical* student takes.

2.2 Skewed Distributions

This example, where a single student took an unusually large number of globalized courses, brings up the important topic of skewed distributions. This student's globalized coursework would increase the mean number of globalized courses taken among students in the dataset even when all the other students took only a few globalized courses. In quantitative analysis, we say that distributions can be **right skewed** (another word you will see in the literature for this kind of distribution is **positively skewed**), meaning that there are only a few values that fall along the right-hand side of the distribution, or **left skewed** (**negatively skewed**), meaning

---

[1] The equations throughout this book apply to *samples* rather than *populations* because, in most situations, international education researchers do not have access to data representing a full population. Interested readers are encouraged to explore a more advanced statistics textbook for equations that are appropriate for population data. That said, when samples are large enough, the numerical difference between sample and population calculations is minimal.

2.2 Skewed Distributions

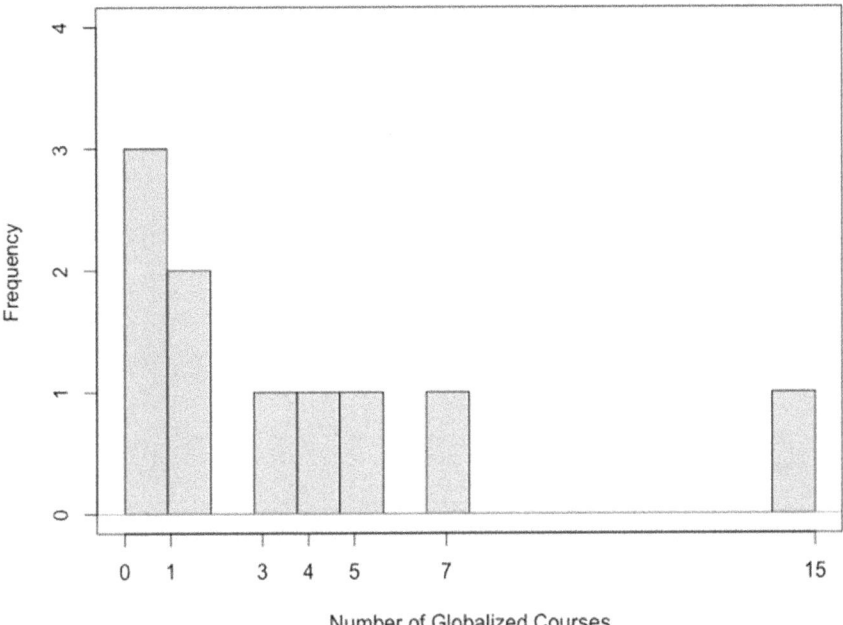

Fig. 2.2 Number of globalized courses taken in students' first year of enrollment and example of a right skewed distribution. (*Note* Data in this figure are invented for illustrative purposes.)

that there are only a few values that fall along the left-hand side of the distribution. The example in Fig. 2.2 is right skewed—only a few students took a large number of globalized courses. The distribution depicted in Fig. 2.3 is an example of a left-skewed distribution—here, only a few students took a low number of globalized courses (perhaps this institution has an initiative that encourages first-year students to take globalized courses). Importantly, distributions can be skewed without the presence of just a few outliers—some variables simply take a skewed shape.

While looking at a visual display of the distribution is helpful in determining whether a distribution is skewed, it is not a necessary step to determine if a distribution is skewed. In a right-skewed distribution, the presence of a few larger values raises mean, so the mean is usually higher than the median (in the example in Fig. 2.2, the mean, 3.6, is higher than the median, 2), while in a left-skewed distribution, a few smaller values lower the mean, so the mean is usually lower than the median (in the example in Fig. 2.3, the median is 11 and the mean is 8.7). In reality, most distributions are skewed to a certain extent, and analysts are more concerned with how much rather than if a variable's distribution is skewed.

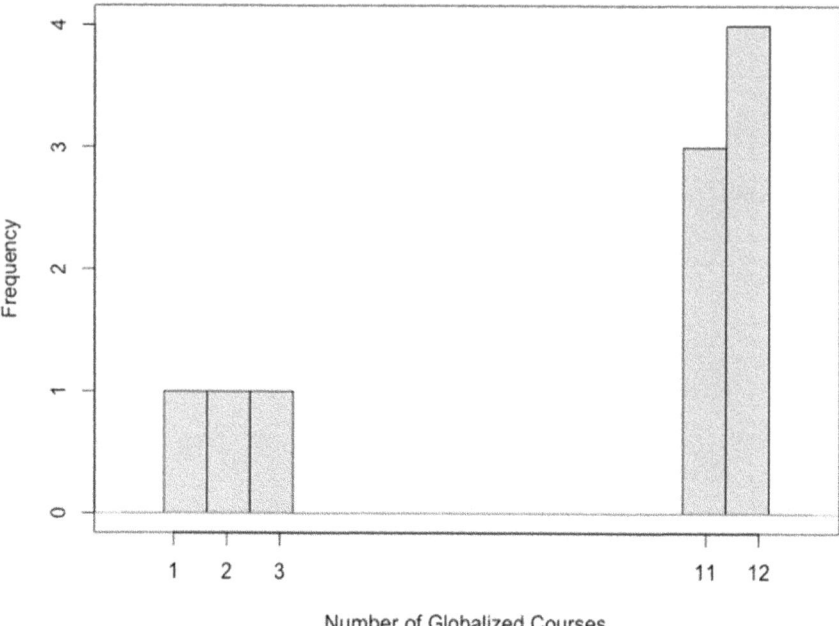

Fig. 2.3 Example of a left skewed distribution. (*Note* Data in this figure are invented for illustrative purposes.)

2.3 The Normal Distribution (Part 1)

This discussion about the skew of a distribution brings us to a specific kind of distribution that is most definitely *not* skewed: the normal distribution. The **normal distribution** is really a hypothetical distribution. In reality, very few variables exactly follow a normal distribution, although it is thought that biological variables, such as human height or weight, are typically normally distributed. While the normal distribution is introduced in this chapter, and we will build on it in the next chapter, its importance in quantitative analysis will really only become apparent when we get to Chap. 4, where we discuss hypothesis testing.

Figure 2.4 illustrates the normal distribution, which follows a general bell shape that you may already be familiar with. Several characteristics of the normal distribution jump out immediately. First, it is **symmetrical**—that is, data are distributed so that the frequency of values on either side of the mean are a reflection of one another (in this example, the mean value is 100). Second, the normal distribution is **unimodal**, meaning that it only has one most frequent value (only one "peak"). This most frequent value is also the mean and the median. That is, the average value is also the value in the middle of the distribution, in addition to being the most frequent value. A final characteristic of the normal distribution is that it is **asymptotic**. This means that the tails of the distribution never quite

2.4 Examples from Sample Data and the Literature

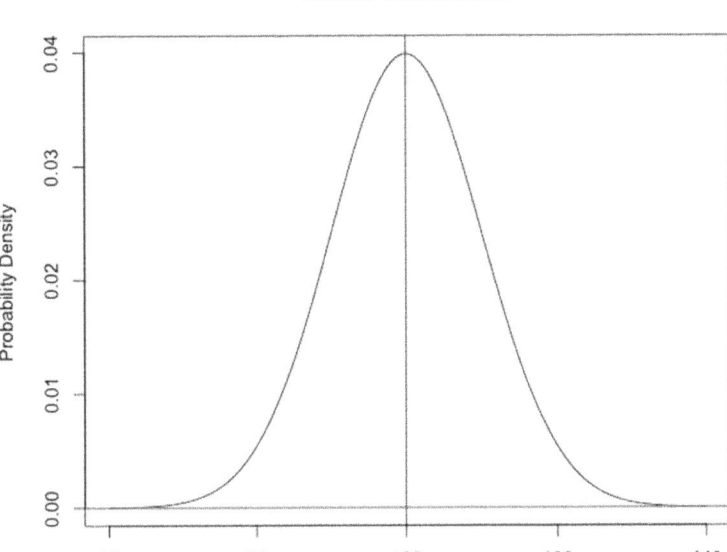

Fig. 2.4 The normal distribution

touch the horizontal axis (the x-axis), although they get very close. Described differently, because the normal distribution is asymptotic, all values have a non-zero probability of occurring along the distribution, even if this probability is very small.

Because of these three properties of the normal distribution (that it is symmetrical, unimodal, and asymptotic), if you were to choose a value from the normal distribution at random, you would be less likely to choose a value further from the mean (but choosing one of these values would still be possible) and more likely to choose a value that falls close to the mean, regardless of whether the value were to the left or the right of the mean. This is not the case for skewed distributions, such as the left-skewed distribution in Fig. 2.3. If we were to select a random student from this distribution, we would be much more likely to select a student who took more than the mean number of globalized courses. That is, in this example, seven students took more than the mean number of globalized courses ($\overline{X} = 8.7$), while only three took fewer than this mean value.

We will leave this discussion of the normal distribution for now but will return to it in the next chapter.

2.4 Examples from Sample Data and the Literature

To illustrate measures of central tendency using real data, we return to the distribution presented at the beginning of this chapter in Fig. 2.1, which illustrates the number of international students studying at U.S. state flagship institutions (each state has one flagship institution, meaning that there are 50 institutions represented in this dataset) in the 2015–16 academic year. In this particular academic year, the mean number of international students enrolled at flagship institutions was 3310, while the median was 2448. There is no mode for this distribution as each of the 50 flagship institutions presents a unique value for the number of enrolled international students. This situation is common when a variable can take many possible values. Without looking back at Fig. 2.1, we know that this distribution is right skewed, meaning that there are only a few high values, because the median is lower than the mean. This kind of information that describes international student enrollment at a particular type of institution may be especially valuable for individuals who work with international students at these institutions, for the purpose of peer comparisons, or for individuals who work in international student recruitment.

Measures of central tendency are a common feature of reports that international education organizations release. For example, the Forum on Education Abroad surveys its members, which include U.S. colleges and universities as well as education abroad program providers (both in the U.S. and outside the U.S.) that serve U.S. students, every few years to produce a *State of the Field* report (see the Forum on Education Abroad's website [https://forumea.org/resources/data-collection/] for more information about this report). The most recent report, published in 2017, provides mean and median statistics for the number of students studying abroad from four categories of Forum members: private U.S. institutions, public U.S. institutions, U.S.-based program providers, and overseas institutions and organizations. These measures of central tendency are provided in Table 2.3. If we compare the mean and median columns for each member group, we see that the mean is higher than the median for private U.S. institutions, U.S. program providers, and overseas institutions and organizations, meaning that these distributions of study abroad participation numbers are right skewed. A few institutions/organizations report higher study abroad participation compared to most other institutions/organizations in the dataset. In contrast, this distribution for public U.S. institutions is left skewed, there are a few institutions in the dataset that report lower study abroad participation than most of the other institutions in the dataset. We know this because the mean is lower than the median. Again, this information might be useful for international educators when benchmarking their own programs against those of other institutions or considering what might be a reasonable increase or decrease in study abroad participation over time.

Table 2.3 Mean and Median U.S. students abroad from forum on education abroad members in the 2016–17 academic year (based on Fig. 2.2 in the 2017 state of the field report)

Member group	Mean	Median
Private U.S. Institutions	354	263
Public U.S. Institutions	331	828
U.S. Program Providers	1327	475
Overseas Institutions and Organizations	427	180

2.5 Practice Problems

Table 2.4 Mode funding sources for internationalization (based on Fig. 2.4 in the 2017 mapping internationalization on U.S. Campuses Report)

	Federal government (%)	State government (%)	Alumni (%)	Private donors (%)	Foundations (%)	Corporations (%)
2006	20	8	18	24	20	7
2011	18	4	15	21	18	6
2016	18	5	28	33	28	10

In another example, the American Council on Education's *Mapping Internationalization on U.S. Campuses* report (see the American Council on Education's website [https://www.acenet.edu/Research-Insights/Pages/Internationalization/Mapping-Internationalization-on-U-S-Campuses.aspx] for more information about this report), which is based on survey data collected every five years from U.S. colleges and universities, relies on measures of central tendency to track changes in internationalization practices over time. This report specifically summarizes mode responses to survey questions. In the most recent report (Helms et al., 2017), for example, the percentage of institutions receiving funding from various sources for internationalization efforts, including the federal government, state government, alumni, private donors other than alumni, foundations, and corporations, is summarized over time. These percentages are reproduced in Table 2.4. Consistently over time, the mode, or most frequent, funding source for U.S. institutions' internationalization efforts was private donors (24% in 2006, 21% in 2011, and 33% in 2016). This information might be especially useful for international educators seeking to diversify funding sources to support their offices and operations.

2.5 Practice Problems

1. The following table contains information about the number of hours per week a group of ten students spent studying for their foreign language class. Use this table to calculate by hand the following pieces of information:
   a. Mode
   b. Median
   c. Mean

Student	Hours studying
1	10
2	15
3	20
4	0
5	3

Student	Hours studying
6	25
7	23
8	9
9	12
10	0

Note Data in this table are invented for illustrative purposes

    d. Is this distribution skewed? If so, in which direction? How do you know?
2. Sample Dataset #1 contains information about the number of students who studied abroad through U.S. flagship institutions in the 2015–16 academic year (studyabroad). Using your statistical software program of choice, find the mode, median, and mean of this variable. Is this distribution skewed? If so, in what direction? How do you know?

Recommended Reading

A Deeper Dive

Urdan, T. C. (2017). Measures of central tendency. In *Statistics in plain English* (pp. 13–20). Routledge.

Note: The Recommended Readings section in Chapter 3 provides additional resources and example studies that make use of measures of central tendency.

Measures of Variability

In the last chapter, we discussed measures of central tendency, which provide three different ways to summarize a variable using a single number: the mode, the median, and the mean. However, you have probably noticed that these measures are lacking in some of the information that we would really like to know about variables. For example, imagine we have data from a survey of international students where they were asked to rate their satisfaction with international student life services at their host institutions. We know that the mean rating was 3.6. To interpret this number, we need some additional information. If we assume ratings were on a scale from one to five, an average of 3.6 could arise from a large group of students providing ratings of 1 and another large group of students providing ratings of 5, with very few students providing ratings in the middle. In this case, the decisions we make based on these data might include directing resources towards addressing the needs of the large group of students that is very dissatisfied with our current services. At the same time, a mean of 3.6 could result from most students providing ratings hovering around 3 or 4, in which case we might focus our attention on initiatives and programs that address international student satisfaction more generally.

Although this is an extreme example, it illustrates that while measures of central tendency are useful, they do not provide all the information we need to know about a variable's distribution. That is, measures of central tendency ignore that different distributions have different shapes. Some distributions are tall and thin (the left panel of Fig. 3.1), with many values hovering around the mean, while others are long and wide (the right panel of Fig. 3.1), with values more distant from the mean. Note that in each of these distributions, the mean is 3.6 (represented with the vertical line in both figures). Of course, other distributions are somewhere in between in terms of shape and can be more complex than these examples.

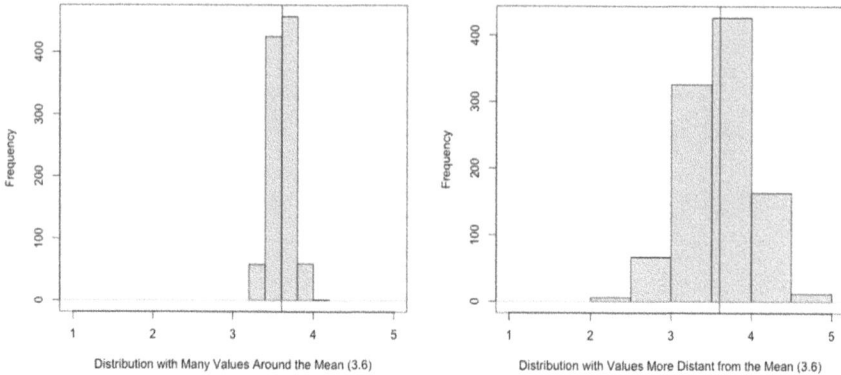

Fig. 3.1 Ratings of international student satisfaction with student life services on a 1–5 scale. (*Note* Data in this figure are invented for illustrative purposes.)

3.1 Range, Variance, and Standard Deviation

Measures of variability, that is, range, variance, and standard deviation, provide a short-hand way of describing the dispersion of values in a distribution.

3.1.1 Range

A distribution's **range** is the maximum value of a variable minus its minimum value. We continue with our example of international student satisfaction ratings to solidify this concept. Table 3.1 contains ratings from a random sample of international students. To calculate the range of this distribution, we take the largest value, 5, and subtract the smallest value, 1, to arrive at 4. Focusing on the range while you are getting to know your data is an important strategy for finding errors or outliers in your dataset. For example, if you know that the survey you distributed asked students to provide ratings on a scale of one to five, but your maximum value is 15, you might suspect that an error occurred somewhere along the way when your dataset was created. You would want to examine the individual observations

Table 3.1 International student ratings of satisfaction with international student services

Student	Rating
1	3
2	5
3	4
4	4
5	1

Note Data in this table are invented for illustrative purposes

3.1 Range, Variance, and Standard Deviation

for these especially high values to ensure that they are accurate and not mistakes. Similarly, you might find that while your maximum value in your dataset is 5, the mean student rating is 1.3. In this case, you might suspect that the student who rated their satisfaction using the 5 category is an outlier.

3.1.2 Variance

While the range is a useful starting point for exploring a variable's distribution, it still tells us nothing about the distribution's shape. That is, is it tall and thin or long and wide? What we need is a measure of the average dispersion from the mean in a distribution. This is exactly what the **variance** and its close relative, the standard deviation, provide. The formula for calculating the variance is as follows:

$$s^2 = \frac{\sum(X_i - \overline{X})^2}{n - 1}$$

Table 3.2 provides an example of how this formula works. To calculate variance (s^2), we first subtract a variable's mean from each value in the dataset ($X_i - \overline{X}$), resulting in a deviation from the mean for each observation (see the third column in Table 3.2). To find a general measure of deviation from the mean, it would seem logical to simply add these deviations up (remember the summation sign, $\sum$, from Chaps. 2) and divide by the number of observations in the dataset (like calculating an average but for deviations). However, because the mean that gets subtracted in this equation is the average value of all the values in a distribution, these deviations add up to zero (go ahead—try it), thus providing us with a fairly useless calculation (zero divided by anything is zero). Instead, we take the square of these deviations before we sum them (see the fourth column in Table 3.2). When we sum these numbers, we get what is called the sum of squared deviations (or the sum of squares). You will see this concept again in later chapters in this book. In our working example, when we add up the squared deviations, we arrive at 9.2 (see the last row in Table 3.2).

Finally, instead of dividing by n to get the average of these squared deviations, the equation for variance tells us to divide by $n - 1$. This is not a mistake. We subtract 1 from n before we divide to account for the fact that we have data from

Table 3.2 Variance calculation in international student ratings of satisfaction with international student services

Student	Rating	$X - \overline{X}$	$(X - \overline{X})^2$
1	3	3 − 3.4 = −0.4	0.16
2	5	5 − 3.4 = 1.6	2.56
3	4	4 − 3.4 = 0.6	0.36
4	4	4 − 3.4 = 0.6	0.36
5	1	1 − 3.4 = −2.4	5.76
Sum			9.20

a sample rather than an entire population (also referred to as adjusting for **degrees of freedom**). In doing so, we assume that the sample mean is different from the population mean and that the variance would be larger if we had access to the population mean to use in our formula. By subtracting 1, we inflate the value of the variance (by dividing by a smaller number), thus accounting for our imprecise estimate of what the population mean actually is.[1] In the case of our example, we divide 9.2 by 4 (there are 5 students in our sample), arriving at a variance of 2.3.

3.1.3 Standard Deviation

You might be wondering how to interpret a variance of 2.3, remembering that when we calculated deviations from the mean, we actually squared them and then used the sum of squared deviations to calculate variance. These squared values are only distantly related to the numbers that we started out with in our distribution (remember, these were international student satisfaction ratings on a scale of 1–5). To arrive at a measure of variability that is more readily interpreted, we often will prefer the **standard deviation** because it is on the same scale as the original variable. To calculate the standard deviation, all we do is take the square root of the variance, illustrated formally in the following equation:

$$s = \sqrt{\frac{\sum (X_i - \overline{X})^2}{n - 1}}$$

In our example, the square root of 2.3 is around 1.52, the standard deviation. The smaller the standard deviation, the more bunched up values are around the mean (and the taller and thinner the distribution). In contrast, the larger the standard deviation, the more spread out values are around the mean (and the wider the distribution). Taken together, the standard deviation and the range provide us with a pretty good picture of what a distribution looks like—even if we are unable to look at the distribution in graphic form.

3.2 The Normal Distribution (Part 2)

In the previous chapter, we introduced the normal distribution, noting that it has several important properties—it is symmetrical, unimodal, and asymptotic. We also

---

[1] A more statistically technical way to describe why we subtract $n - 1$ in the denominator of the variance formula is to appeal to the concept of *degrees of freedom*, as mentioned briefly in the main text of this chapter. Because we use the sample mean to estimate the variance rather than the population mean, we do not have n pieces of information (degrees of freedom) to calculate variance, but rather $n - 1$. That is, if we know the numerical values of $n - 1$ observations and we know the mean, we can calculate the value of the nth observation. This value is not free to vary like the other values, and so we account for that fact in the formula.

3.2 The Normal Distribution (Part 2)

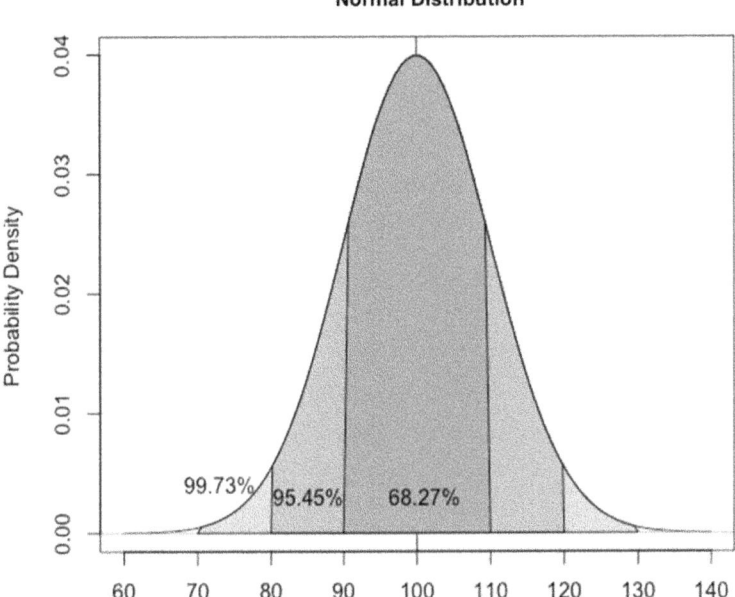

Fig. 3.2 Percentages of the normal distribution within one, two, and three standard deviations from the mean (mean = 100, SD = 10)

know quite a bit about how much it varies. An additional property of the normal distribution is that 68.27% of observations fall within one standard deviation of the mean. This percentage is illustrated in the dark grey portion of the distribution in Fig. 3.2. Moreover, 95.45% of observations fall within two standard deviations of the mean (the medium gray portion of Fig. 3.2 plus the dark grey portion), and 99.73% of observations fall within three standard deviations of the mean (all of the grey shaded portions of Fig. 3.2). This is an important property of the normal distribution that we will exploit in the next chapter when we turn our attention to hypothesis testing. Because we know these percentages, we will be able to calculate exactly what percentage of the standard normal distribution falls above or below a specific value. These percentages represent the probability that we would select an observation at random that is higher or lower than that value. For example, if we were to select a number at random from the distribution in Fig. 3.2, which has a mean of 100 and a standard deviation of 10, there is a 68.27% chance that we would select an observation that is between 90 and 110 (within one standard deviation on either side of the mean).

3.3 Examples from Sample Data and the Literature

To illustrate measures of variability using real data, we return to our example from the last chapter, the number of international students studying at U.S. flagship institutions in the 2015–16 academic year. This distribution is illustrated visually in Fig. 3.3.

Measures of variability for this distribution are provided in Table 3.3. This table indicates that the range of the number of international students studying at U.S. flagship institutions was 10,767 (the maximum value of international students was 11,133 and the minimum value was 366), indicating that there were substantial

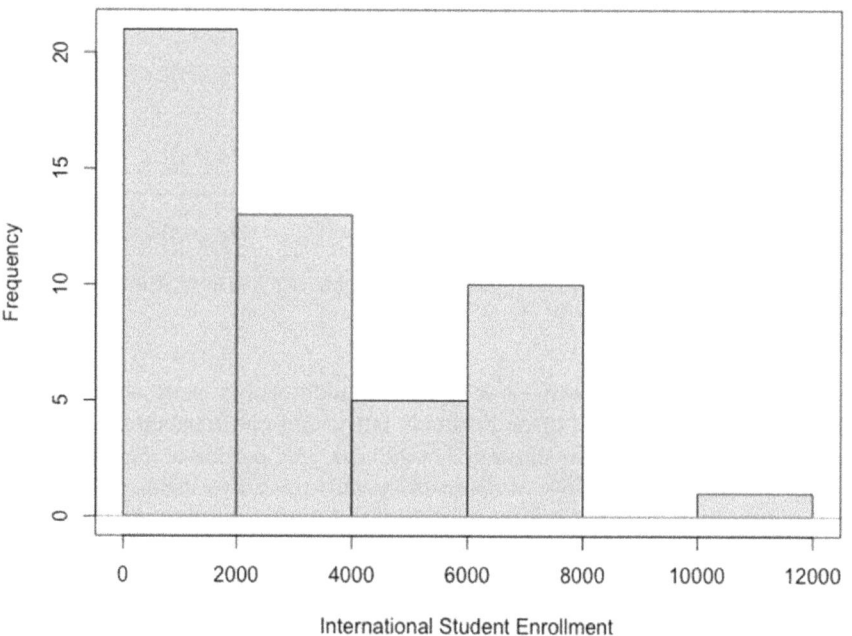

Fig. 3.3 Histogram summarizing international student enrollment at U.S. Flagship Institutions from the 2015–16 Academic Year. (*Data source:* Sample dataset #1 - US National Center for Education Statistics, Integrated Postsecondary Education Data System)

Table 3.3 Measures of variability for the number of international students studying at U.S. Flagship Institutions in the 2015–16 academic year

Range	Variance	Standard deviation
10,767 (maximum value: 11,133, minimum value: 366)	6,331,934	2516

Data source: Sample dataset #1 - US National Center for Education Statistics, Integrated Postsecondary Education Data System

3.4 Practice Problems

Table 3.4 Mean and standard deviation PISA 2018 reading performance scores for five OECD countries (adapted from Table I.B1.4 in OECD [2019])

Country	Mean	Standard deviation
Australia	503	109
Austria	484	99
Belgium	493	103
Canada	520	100
Chile	452	92

differences in the number of international students served at these institutions during this academic year. The variance for this distribution is 6,331,934. When we take the square root of the variance, we arrive at 2516 students, providing us with a better idea of how far away from the mean (which was 3310 students) observations in this distribution typically are. For practical purposes, the range of this distribution might be the most immediately useful for international educators, as it adds nuance to comparisons among institutions. However, the variance and standard deviation are vitally important to additional statistics that we will calculate in future chapters.

Like measures of central tendency, measures of variability, especially standard deviation, are often used in reports produced by organizations with a focus on international education. One such organization is the Organization for Economic Co-operation and Development (OECD), which administers the Programme for International Student Assessment (PISA) to assess 15-year-old students' knowledge of reading, mathematics, and science internationally. Table 3.4 provides a snapshot of students' scores for five countries (the first five in the list of OECD countries alphabetically) in Reading in 2018, taken from Volume 1 of the OECD's report on data collected in this year (see OECD, 2019 for additional information). The standard deviations for test scores from these five countries indicated that in some countries, reading scores exhibit more variation than in others. For example, the standard deviation for reading test scores in Chile was rather low ($s = 92$), while in Australia, the standard deviation was higher ($s = 109$). It is interesting to consider why these differences in variability may occur—a useful endeavor might be to attempt to find patterns in the data, where some country characteristics lend themselves to more variation compared to others.

3.4 Practice Problems

1. You began working with the following table in the previous chapter, which contains information about the number of hours per week a group of ten students spent studying for their foreign language class.

Student	Hours studying
1	10

Student	Hours studying
2	15
3	20
4	0
5	3
6	25
7	23
8	9
9	12
10	0

Note Data in this table are invented for illustrative purposes

You have already calculated measures of central tendency for this dataset by hand. Now, calculate the following pieces of information:

a. Range
b. Variance
c. Standard Deviation.

2. We began working with Sample Dataset #1 in the previous chapter, taking a look at measures of central tendency for the number of students who studied abroad through U.S. flagship institutions in the 2015–16 academic year (study abroad). Using your statistical software program of choice, now find the range, variance, and standard deviation of this variable. If your software has plotting functions, plot the distribution to visualize this variable.

Recommended Reading

A Deeper Dive

Agresti, A., & Finlay, B. (2009). Descriptive statistics. In *Statistical methods for the social sciences* (pp. 31–72). Pearson.
Urdan, T. C. (2017a). Measures of variability. *Statistics in plain English* (pp. 21–32). Routledge.
Urdan, T. C. (2017b). The normal distribution. *Statistics in plain English* (pp. 33–42). Routledge.
Wheelan, C. (2013). Descriptive statistics: Who was the best baseball player of all time? *Naked statistics: Stripping the dread from data* (pp. 15–35). Norton.

Additional Examples

Ibrahim, A. (2018). The happiness of undergraduate students at one university in the United Arab Emirates. *International Journal of Research Studies in Education, 7*(3), 49–61.

Park, E. (2019). Issues of international students' academic adaptation in the ESL writing class: A mixed-methods study. *Journal of International Students, 6*(4), 887–904.

Hypothesis Testing

The two preceding chapters introduced a distribution that is very important in statistics—the normal distribution (depicted here in Fig. 4.1). Chapter 2 introduced the normal distribution as a hypothetical distribution with several important properties. First, the normal distribution is symmetrical, meaning that data are distributed so that the frequency of values on either side of the mean reflect one another. A second property is that the normal distribution is unimodal, meaning that it only has one most frequent value (the mode). This value is also the mean and the median. Third, the normal distribution is asymptotic, meaning that the tails of the distribution never quite touch the horizontal axis, even if they do get very close. In other words, all values along the horizontal axis have a non-zero probability of occurring, even if this probability is quite small.

Chapter 3 introduced a fourth property of the normal distribution—namely that we know what percentage of observations fall within one, two, and three standard deviations of the mean when a variable is normally distributed. Another way of looking at this property is that the shape of the normal distribution is determined by only two things: the mean and the standard deviation. We know that 68.27% of observations fall within one standard deviation from the mean of a normally distributed variable, 95.45% fall within two standard deviations, and 99.73% fall within three standard deviations. Because we know these percentages, we can determine the probability that a specific value falls within one, two, or three standard deviations from a distribution's mean. For example, in Fig. 4.1, the mean of the distribution is 100 while the standard deviation is 10. In this example, if we were to draw a random value from the distribution, the probability that this value would fall between 70 and 130, within three standard deviations from the mean on either side, is 99.73%. The probability that this value would fall between 80 and 120, within two standard deviations from the mean, is 95.45%. The probability that this value would fall between 90 and 110, within one standard deviation on either side of the mean, is 68.27%. We can determine these probabilities because of the special properties of the normal distribution: symmetry, unimodality, its asymptotic

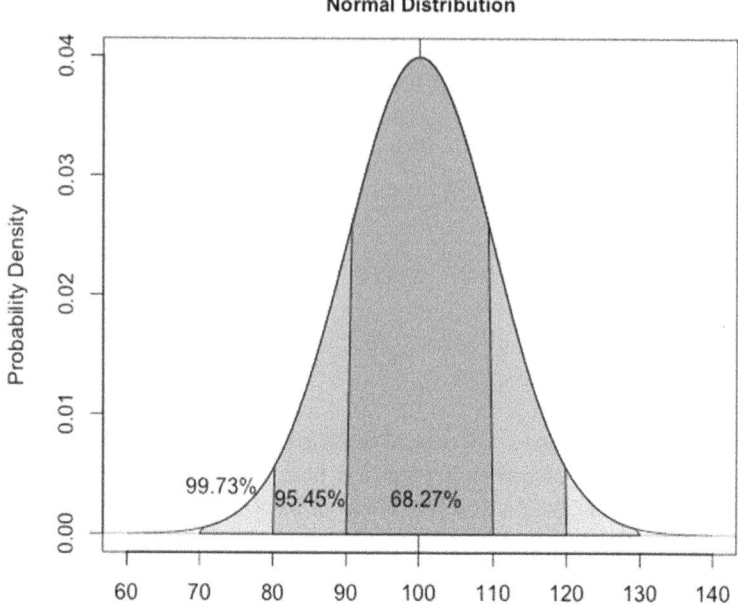

Fig. 4.1 The normal distribution (mean = 100, SD = 10)

character, and the known distribution of observations. These properties are key to hypothesis testing and, for that matter, most of the statistics that you will see in this book. It is worth noting that a standardized version of the normal distribution, called the z-distribution, rescales continuous variables so that the mean is set to 0 and each standard deviation is equal to 1. This is common practice in statistics.

As discussed in Chap. 1, quantitative researchers often test hypotheses about a particular population of interest using data from a sample of individuals from that population rather than the entire population. That is, we use sample data to make inferences, or draw conclusions, about the population. An example will help to make this idea concrete. Consider that you serve on an institutional committee that oversees the conversion of students' grades from an international institution's grading system to the US's system, which assigns students a grade point average (GPA) on a scale that ranges from 0.0 to 4.0, with 0.0 corresponding to the lowest and 4.0 corresponding to the highest grade possible. Senior leadership at your institution is concerned that international grade conversion gives unfair advantage to study abroad students. You have been asked to determine whether the mean term GPA of the population of returning study abroad students is meaningfully different from the mean term GPA of all students, which at your institution is 3.0. To explore this possibility, you would ideally explore the academic records of all returning study abroad students, that is, the full population. However, due to time and resource constraints, you are only able to collect term academic records data from a randomly-selected sample of 50 study abroad returnees. You will have to

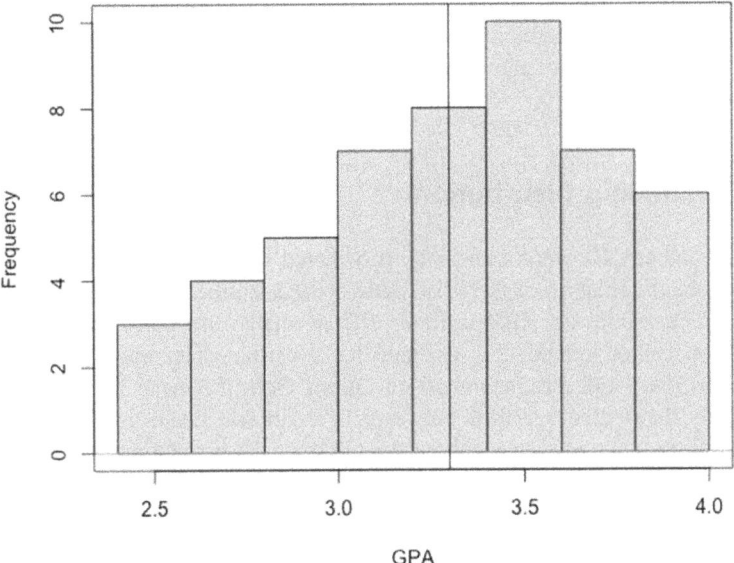

Fig. 4.2 GPA distribution for 50 randomly-selected study abroad returnees (mean = 3.3, SD = 0.41) (*Note* Data are invented for illustrative purposes.)

use this sample to draw conclusions about the entire population of study abroad returnees. The distribution of GPA in your sample is illustrated in Fig. 4.2, where the mean GPA is 3.3, indicated with the vertical line, and the standard deviation is 0.41.

Since the mean GPA in your sample is above your hypothesized population mean of 3.0, you might be tempted to assume that grade conversions benefit returning study abroad students. However, this conclusion would be premature if not incorrect. Notice that several students in your sample do have GPAs that are below 3.0, and there is a chance that this sample includes a disproportionate number of students with high grades. If you were to take a random sample of 50 returning study abroad students again, you might end up with a sample mean GPA of 2.8 or 1.7. In other words, we do not know if this specific random sample is truly representative of the entire population of study abroad returnees. To make inferences about the mean GPA of entire population of returning study abroad students we need to account for the fact that every time we take a random sample from the population of study abroad returnees, the mean sample GPA will be different. A starting point for making inferences about a population is to perform a hypothesis test, a statistical approach that assesses whether a sample mean is meaningfully different from a hypothetical (or hypothesized) population mean.

The question that will guide our exploration of hypothesis testing is the following: What is the probability of obtaining a sample of returning study abroad students with a mean GPA of 3.3 if the population mean is truly 3.0? In other words, I am interested in whether the difference between my sample mean and the hypothesized population mean is meaningfully different or if this difference is more likely due to random chance. In the first part of this chapter, I outline how we can make use of the properties of the normal distribution to answer this question.

4.1 Sampling Distribution

If we knew that GPA were a normally distributed variable, we could infer quite a bit about the distribution of GPAs of study abroad returnees based on our sample. In real life, variables like GPA are not often normally distributed in a population, and so we cannot immediately assume that a certain percentage of GPAs falls within one, two, and three standard deviations from the mean like we can with a normally distributed variable. However, we can still make use of the normal distribution to make inferences about the GPAs of the population of study abroad returnees based on our sample. To explain how the normal distribution is useful even in these contexts, we turn to a few concepts from probability theory that are somewhat abstract. While these concepts may not seem especially practical at first, we will return to their implications for our GPA question later in this chapter.

We begin with a foundational theorem in statistics, the Central Limit Theorem.[1] The Central Limit Theorem helps us to make use of the properties of the normal distribution even when we are dealing with variables that are not normally distributed, like our GPA example. Essentially, what the Central Limit Theorem shows is that even when we are working with a variable that is not normally distributed, the properties of the normal distribution, such as its symmetry and distribution of observations by standard deviation, still apply as long as we are working with a **sampling distribution of the mean** (or *sampling distribution* for short) that makes use of large enough samples. To obtain a sampling distribution, we draw multiple random samples from a population and record the mean value of our variable of interest each time, essentially creating a new dataset of calculated sample means. What the Central Limit Theorem shows is that, even in cases when a variable itself is not normally distributed, the sampling distribution will approximate a normal distribution, so long as each sample is large enough (the rule of thumb is 30, but the Central Limit Theorem can also hold even when samples are smaller).

In our running example, a sampling distribution is what we would obtain if we were to draw a random number of students from the population of study abroad returnees, find the mean of their GPAs, and record this number in a new dataset. We would perform this process again and again, calculating and recording the sample mean each time. With large enough samples, the distribution of these sample means would eventually approximate a normal distribution. Fig. 4.3 illustrates

---

[1] The full details of the Central Limit Theorem are beyond the scope of this book, but I encourage you to explore them further in a more advanced statistics textbook, such as those listed in the recommended reading list at the end of this chapter.

4.1 Sampling Distribution

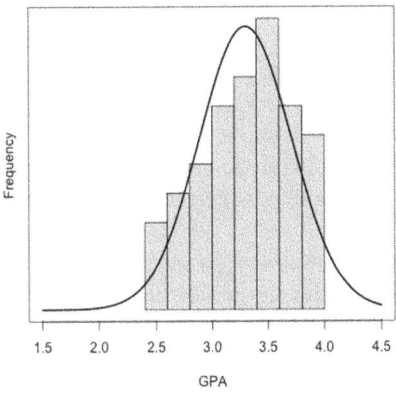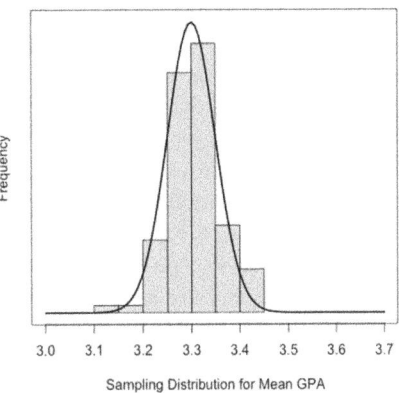

Fig. 4.3 Distribution of GPA (left) and sampling distribution of mean GPA (right) with the normal distribution superimposed (black line). (*Note* Data are invented for illustrative purposes.)

this concept. The plot on the left side is the same GPA distribution illustrated in Fig. 4.2 from our randomly selected sample of 50 returning study abroad students. The curve superimposed on top of this distribution represents the normal distribution. While this GPA distribution is not entirely different from the normal distribution, it is a bit left skewed. That is, some students have GPAs that are higher than what we would expect if GPA were a normally distributed variable.

The plot on the right side of Fig. 4.3 depicts a sampling distribution for GPA. To construct this sampling distribution, I took advantage of data that we are unable to obtain in the reality of our running example—the GPAs of all returning study abroad students. To construct this sampling distribution, I took 100 random samples of 30 students each from this population of study abroad returnees and recorded the mean GPA for each sample in a new dataset. This plot is the distribution of these sample means. As you can see from the normal curve superimposed on top of this sampling distribution, this distribution is a better representation of the normal distribution. In both plots in Fig. 4.3, the mean value is 3.3; however, the plot to the right more closely approximates the normal distribution.

4.1.1 Expected Value of the Mean and Standard Error

Two additional pieces of information are useful for when we talk about sampling distributions. First, like the individual variable distributions that we saw in Chaps. 2 and 3, sampling distributions have a measure of central tendency, the **expected value of the mean** (or *expected value* for short), and a measure of variability, the **standard error**. The expected value of the mean is a relatively straightforward measurement—it is the mean of the sampling distribution of the mean. An important property of the sampling distribution is that its expected value is, by definition, equal to the population mean. In our running example, we will hypothesize that the population mean is 3.0 based on the mean term GPA of all students, but as the right-hand plot in Fig. 4.3 indicates, the true population mean is 3.3. In terms of variability, like standard deviation, the standard error of a sampling distribution provides us with information about the variability of a sampling distribution. The larger the standard error, the more variation there is among sample means. The greater this variation, the less confident we can be that any one sample mean is representative of the population. The standard error in the sampling distribution in Fig. 4.3 is 0.05.

At this point, you might be wondering how these very abstract concepts are useful in real life. Only rarely do education researchers ever have the opportunity to take multiple samples from the same population. Moreover, doing so is often logistically impractical and, even if possible, could become very expensive very quickly (whether in terms of money or time). Normally, we can only gather one sample to test a hypothesis. So how do we calculate the expected value and standard error of a hypothetical sampling distribution that we are unlikely to ever have the data to construct in real life? In the case of the expected value of the sampling distribution, this statistic is what we use to form hypotheses. In our running example, we established that we want to test the hypothesis that the mean GPA for all study abroad returnees is not meaningfully different from 3.0, so our hypothesized expected value is 3.0 (even though we know this isn't the case, given the sampling distribution represented in the right-hand plot in Fig.4.3. Remember that we do not in reality have access to this information).

When we work with samples, in contexts like our running example where we do not have access to population data, we rely on the sample standard deviation to estimate the standard error of the sampling distribution. Specifically, the estimated standard error is calculated as follows:

$$s_{\bar{x}} = \frac{s}{\sqrt{n}}$$

In this equation, the estimated standard error ($s_{\bar{x}}$) is the sample standard deviation (s, see Chap. 3 on how to calculate the standard deviation of a variable) divided by the square root of the sample size ($\sqrt{n}$). We take the square root of the sample size in the denominator because otherwise researchers could manipulate sample size to make the standard error very small. That is, if we simply divided by sample size in this equation, the larger the sample size, the smaller the standard error would be, regardless of how much variation is actually present in a sample. Taking the square root of the sample size in the denominator means that there are diminishing returns on increasing sample size when estimating the standard error. In our running example, the standard error would be calculated as follows, using the sample standard deviation of 0.41 in the numerator:

$$s_{\bar{x}} = \frac{0.41}{\sqrt{50}} = 0.06$$

This calculation provides the best estimate of the standard error that we can obtain from a single sample. We will need this information later when we perform a hypothesis test.

Notice that with the sample size in the denominator, the standard error is, by definition, smaller as sample size increases, in spite of the diminishing returns on sample size just noted. The rationale behind this property of the standard error is intuitive in that larger samples include more members of a sample's population of interest, and so extreme values in the sample have less influence on the sample mean. Assuming random selection, the more members of a population in a sample, the less guesswork, or error, there is between the sample mean and the population mean. In other words, the larger your sample size, the closer your estimate of the mean is likely to be to the true population mean.

4.1.2 The *t*-Distribution

Clearly, sample size is important in that it helps us estimate the standard error of a sampling distribution using the standard deviation of a single sample. In the context of a sampling distribution, the standard error is used in place of the standard deviation to determine the segments of a distribution that fall within 68.27, 95.45, and 99.73% of the expected value of the mean, which is also our hypothesized population mean (see Fig. 4.1). If the standard error is estimated inaccurately or with substantial uncertainty, as would be the case with a very small sample, then the Central Limit Theorem falls apart, and we can no longer rely on its properties for the purpose of approximating a normal distribution and subsequent hypothesis testing. In other words, we would no longer be able to make useful assumptions about the probability that a particular sample mean falls a certain distance from a hypothesized mean in the sampling distribution. In our running example, this is exactly our concern—is the sample mean of 3.3 a meaningful distance (as measured by the estimated standard error) from the hypothesized population mean of 3.0?

One caveat that we should note is that because the estimated standard error calculation depends on sample size, when sample size is small, the estimated standard error will be large regardless of how closely it approximates the true population. With a larger sample, the estimated standard error will be small, again, regardless of its accuracy. This happens even though we have already accounted for this potential issue by taking the square root of the sample size in the denominator of the estimated standard error calculation. Is there any way that we can account for this pitfall and still use the properties of the normal distribution to test hypotheses? Luckily, we can use what is called a *t*-distribution rather than the normal distribution to account for this problem whenever we use a sample standard deviation to estimate the standard error. The *t*-distribution looks almost identical to the normal distribution, but the edges of the distribution, referred to as *tails*, are thicker than the normal distribution and vary in just how thick they are depending on sample size (technically, these are degrees of freedom). When the tails of the

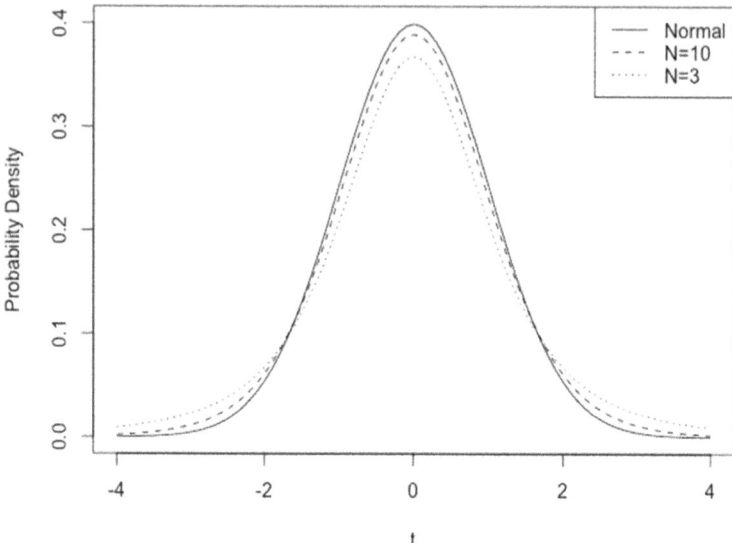

Fig. 4.4 The *t*-distribution

t-distribution are relatively thick, the percentage of observations falling within one, two, or three standard deviations of the mean (the middle of the distribution) is smaller, thus accounting for uncertainties regarding sample size and the estimated standard error.

An important property of the *t*-distribution is that, as sample size gets larger, it becomes indistinguishable from the normal distribution, and so any estimate of the percentage of observations within one, two, or three standard deviations from the mean is the same as if we were using the normal distribution. The *t*-distribution is also a standardized distribution, meaning that the mean always equals 0 and the standard deviation always equals 1. Examples of the *t*-distribution with different sample sizes are shown in Fig. 4.4. Notice that the tails of the *t*-distribution when sample size is 3 (the dotted line) are thicker than the tails of the *t*-distribution when sample size is 10 (the dashed line). For comparison, the solid black line in this figure represents the normal distribution.

We now have everything we need to address the question that we posed at the beginning of this chapter: What is the probability of obtaining a sample of returning study abroad students with a mean GPA of 3.3 if the population mean is truly 3.0?

4.2 Hypothesis Testing with One Sample

A useful way to think about our research question for the purpose of hypothesis testing is to turn it into two hypotheses: a **null hypothesis** (H_0) and an **alternative hypothesis** (H_A) as follows:

H_0: The mean GPA of study abroad returnees is 3.0.
H_A: The mean GPA of study abroad returnees is not 3.0.

The language that I used to outline these two hypotheses is relatively standard in statistics. By convention, the null hypothesis is always that there is no significant difference between two values while the alternative hypothesis is always that there is one. Since we do not know the true population mean GPA, we play it safe and assume that the null hypothesis is correct (that the population mean is 3.0) and then calculate the probability of getting our sample mean (3.3) or a more extreme one if the null hypothesis remains true. In other words, we use our hypothesized population mean GPA, which is also the expected value of the mean, and the estimated standard error, to construct a hypothetical sampling distribution. We then use this sampling distribution, which we know is normally distributed, to calculate the probability that a random sample would yield a mean GPA of 3.3. As this probability becomes smaller, we determine that either the null hypothesis remains true, and our sample is a poor representation of the population, or that there is nothing wrong with our sample and the null is false.

These null and alternative hypotheses are examples of two-tailed hypotheses, meaning that they refer to both the high and low ends of the distribution of GPAs of study abroad students. A **two-tailed hypothesis test** refers to both tails of the *t*-distribution. Notice that there are two ways in which a sample mean GPA could be different from a hypothesized population mean GPA of 3.0: it could be higher (e.g., 3.8) or it could be lower (e.g., 2.3). We can also test hypotheses that refer to only one tail of the distribution, called a **one-tailed hypothesis test**. We might test a one-tailed hypothesis test if we are specifically concerned about study abroad returnees being assigned unfairly *higher* grades than other students. Examples of null and alternative hypotheses for a one-tailed test are:

H_0: The mean GPA of study abroad returnees is not greater than 3.0.
H_A: The mean GPA of study abroad returnees is greater than 3.0.

In this case, these two hypotheses focus attention only on one tail of the *t*-distribution, the high end (note that other hypotheses might focus on the lower end of the distribution). We will move forward with these two sets of hypotheses in this section.

4.2.1 One-Sample *t*-tests

In this formula, the numerator is the difference between the sample mean ($\overline{X}$) and the hypothesized population mean (μ_0) while the denominator is the standard error ($s_{\overline{x}}$) that we calculated earlier in this chapter. The numerator in this formula is the comparison that we are interested in—the difference between our sample mean and the proposed population mean. The standard error in the denominator accounts for dispersion in the sampling distribution. If dispersion in the sampling distribution is quite high, then obtaining a sample mean that is substantially higher or lower than the population mean has a high probability. In this case, we would not be able to say that our sample mean was meaningfully different from the hypothesized population mean, regardless of the measured difference between the two mean values. In contrast, if dispersion in the sampling distribution is low and we have a sample mean that is substantially different from the population mean, then our sample mean is likely significantly different from the hypothesized population mean.

Throughout this chapter, I have noted the pieces of information that we need to calculate a t-statistic for our running example. The sample mean is 3.3 while the hypothesized population mean is 3.0. The standard deviation for my sample of 50 students is 0.41. We can use this information to calculate t. First, remember that when we calculated the standard error previously we arrived at the following:

$$s_{\bar{x}} = \frac{s}{\sqrt{n}} = \frac{0.41}{\sqrt{50}} = 0.06$$

The result of this calculation is the denominator of the formula for t. I use this standard error along with the sample and population means to calculate t:

$$t = \frac{\overline{X} - \mu_0}{s_{\bar{x}}} = \frac{3.3 - 3.0}{0.06} = 5.00$$

Now that I have calculated t, the next step is to refer to the t-distribution. In the case of a two-tailed test, I determine the percentage of the area under the distribution curve that is greater than 5 or less than -5 (remember that in two-tailed tests, we are not concerned with whether the sample mean is higher or lower than the hypothesized mean, only that it is different). In the case of a one-tailed test, I determine just the percentage of this area that is greater than 5. The latter percentage represents the probability that I would obtain a sample with a mean GPA of 3.3 *or higher* if the population mean GPA was, in fact, 3.0.

Calculating the area under a curve like the t-distribution is a complex task made even more complicated by the fact that the shape of the t-distribution is adjusted depending on sample size. Luckily, we do not have to make these calculations ourselves since the t-distribution is standardized and is thus the same every time. Appendix A provides what are called **critical values** in the t-distribution. A critical value is a value in a distribution (in this case, the t-distribution) that is used as a reference point in hypothesis testing. Critical values mark points along a distribution that are associated with specific percentages of area under the curve. These percentages are then associated with probabilities that a specific sample mean will be derived from a hypothesized population. Notice that the first column in this table in Appendix A refers to degrees of freedom (df). This is where we find the t-values we need to use based on our sample size. Like we did in Chap. 3 when we calculated variance (and for the same reason), we subtract 1 from our sample

4.2 Hypothesis Testing with One Sample

size $(n - 1)$ and find the row associated with the appropriate degrees of freedom for our sample. Notice that as our sample size gets larger, subtracting 1 matters less and less. That is, as sample size gets larger, the *t*-distribution approximates the normal distribution, and the adjustment for sample size is no longer especially important. In the case of a large sample size, we would use the last row of this table (infinity).

Next, notice that at the top of this table, there are two options: one-tailed and two-tailed tests. We select one row or the other depending on the nature of our hypotheses. These two rows additionally make reference to **alpha levels** (identified with the Greek letter α). This part of hypothesis testing is left up to the discretion of the researcher. Researchers choose the appropriate alpha level to test their hypotheses and use this alpha to determine whether the difference between means (in this case, a sample mean and a population mean) is statistically significant. In education research (and the social sciences more generally), alpha is typically set to 0.05. For the time being, we will set $\alpha = 0.05$ without worrying too much about what this number means. I return to additional details about alpha later in this chapter.

Once we have found the row associated with the appropriate degrees of freedom for our sample, determined whether our hypotheses are one-tailed or two-tailed, and found the column associated with our selected alpha level, we compare our calculated *t*-value (in our case, 5.00) to the critical *t*-value in the table. (Note that in a two-tailed test, we use the absolute value of the calculated *t*-value because we are concerned with both tails of the distribution, one on either side of the standardized mean of zero. For example, if $t = -5.00$, we use $|t| = 5.00$.) Remember that critical *t*-values are associated with specific percentages of a distribution that fall under the curve of the normal distribution. They provide us with information about the probability that our sample is drawn from a hypothesized population. If our calculated *t*-value is larger than the critical *t*-value, we reject the null hypothesis and conclude that our sample of study abroad returnees is not drawn from a population with a mean GPA of 3.0, our hypothesized value.[2] If our calculated *t*-value is smaller than the critical *t*-value, we fail to reject the null hypothesis, and determine that our sample offers no proof that the true population GPA of study abroad returnees is different from 3.0.

In this example, first, we explore our two-tailed hypotheses:

H_0: The mean GPA of study abroad returnees is 3.0.

H_A: The mean GPA of study abroad returnees is not 3.0.

From the table in Appendix A, I first search for the row corresponding to df = 49, which corresponds to my sample size of 50 (50 − 1 = 49 degrees of freedom). Notice that we do not have a df = 49 row, so we choose the next *lowest* row of df = 40, as this is the conservative choice. Then, I find the number in this row that corresponds to the column for a two-tailed test at $\alpha = 0.05$. In this case, the critical *t*-value is 2.02. Because the absolute value of my calculated *t*-value,

---

[2] Note here that we also assume that our sample is valid.

5.00, is greater than 2.02, I reject the null hypothesis and conclude that the mean GPA of study abroad returnees is not 3.0, the hypothesized population mean (note here that I make no assumptions about the *direction* of the difference between the sample mean and the hypothesized population mean—whether one is greater or less than the other). Note that if we instead had calculated $t = -5.00$, we would have drawn the same conclusion because $|t| = 5.00$.

We follow a similar process for the one-tailed hypothesis test:

H_0: The mean GPA of study abroad returnees is not greater than 3.0.
H_A: The mean GPA of study abroad returnees is greater than 3.0.

In the same row for 40 degrees of freedom in the table in Appendix A (again, going for the next row *lower* than 49), we find the critical *t*-value corresponding to $\alpha = 0.05$ for a one-tailed *t*-test, 1.68. Because the calculated *t*-value, 5.00, is greater than 1.68, I again reject the null hypothesis and conclude that the mean GPA of study abroad returnees is greater than 3.0, the hypothesized population mean (note that this time, I attribute a direction to the difference between the two means).

An interesting issue emerges in one-tailed hypothesis tests when the calculated *t*-value is negative. Our null and alternative hypotheses could be something like the following (notice that I have simply replaced the word 'greater' with 'less' from the hypotheses we just tested):

H_0: The mean GPA of study abroad returnees is not less than 3.0.
H_A: The mean GPA of study abroad returnees is less than 3.0.

To illustrate how this hypothesis test works, let us pretend that our sample mean is lower—say, $\overline{X} = 2.8$—than our hypothesized mean of 3.0:

$$t = \frac{\overline{X} - \mu_0}{s_{\overline{X}}} = \frac{2.8 - 3.0}{0.06} = -3.33$$

However, when you look at Appendix A, all the critical *t*-values are positive. What do you do when calculated *t* is negative? In this case, we take advantage of one of the key properties of the normal distribution (remember that the *t*-distribution is an approximation of the normal distribution): it is symmetrical. While all the critical *t*-values in Appendix A refer to the right-hand side of the *t*-distribution—the side of the distribution that is *above* the standardized mean of 0—their negative values correspond to the left-hand side of the distribution, the side of the distribution that is *lower* than 0. For this reason, we can use the absolute value of our calculated *t* to compare with the critical *t*-values in Appendix A: $|-3.33| = 3.33$. Since 3.33 is larger than the critical *t*-value of 1.68, we reject the null hypothesis and conclude that our sample mean GPA of 2.8 is significantly lower than the hypothesized mean of 3.0.

Alpha. Remember previously that I introduced the idea of an alpha level, a part of hypothesis testing that is left up to the discretion of the researcher. Here, I provide more information about what alpha is and why it matters in hypothesis testing. The alpha level that we just used to evaluate our hypotheses, 0.05, is a standard alpha level in education research. To understand what alpha means, it is

4.2 Hypothesis Testing with One Sample

important to remember that we are dealing with probabilities rather than certainties. That is, while we determined that it is very probable that our sample mean GPA of 3.3 is significantly different from the hypothesized population mean, 3.0, we allow some room for *chance* or *sampling error*, meaning that when I randomly sampled students' GPAs, I might have ended up with a bad sample that is not representative of the population of study abroad returnees. By selecting an alpha level that is 0.05, I am saying that I am willing to conclude wrongly that there is a significant difference between my sample and hypothesized means in 5% of occasions when there really is no significant difference. That is, I am willing for my hypothesis test to be an inaccurate assessment of reality, rejecting the null hypothesis when I should not, 5% of the time. Hypothesis testing is set up in this way so that a researcher needs very strong evidence to reject the null hypothesis. If I am especially concerned about error in my hypothesis testing, I might choose a stricter alpha level as my criterion for determining if a difference in means is statistically significant. Other alpha levels that are common in education research include 0.01 (I am willing to be wrong 1% of the time) and 0.001 (I am willing to be wrong 0.1% of the time).

In reality, there are two kinds of errors that we can make when we conduct a hypothesis test. **Type 1 error**, the kind of error I just described when explaining alpha, happens when I reject the null hypothesis when I should not. In our example, a Type 1 error would occur if the mean GPAs (sample and hypothesized) are not actually different from one another, but I detected a difference in testing the null hypothesis. **Type 2 error** is another kind of error that can occur in hypothesis testing. This kind of error happens when I fail to reject the null hypothesis when in reality I should. In our example, Type 2 error would occur if the mean GPAs of 3.3 and 3.0 were significantly different from one another, but I did not detect this difference in my hypothesis test.[3] Early in this chapter, when we were first considering the difference between the sample mean GPA of 3.3 and the hypothesized mean GPA of 3.0, if we had determined that a difference of 0.3 was not especially meaningful, we would have been committing at Type 2 error. Alpha refers to the Type 1 error rate, which usually gets more attention because falsely rejecting a null hypothesis is generally thought to be a bigger problem to avoid in research.

***P*-values**. In addition to determining a cut-off point for statistical significance, alpha is also directly related the ***p*-value** associated with a particular hypothesis test. The *p*-value (*p* for *probability*) represents the probability that the null hypothesis is correct. If alpha is set to *0.05* and we reject the null hypothesis, then $p < 0.05$, meaning that the probability that we would obtain a sample mean of 3.3

---

[3] When I teach Type 1 and Type 2 error, I often draw a parallel to my dogs, Taca, Bernice, Lola, and Fritsi, barking at the front door of my house. Sometimes, they bark at the front door even when no one is there (perhaps there are ghosts in my neighborhood?). This is a Type 1 error. They detect an effect (effect = someone at the door) that does not exist in reality. Other times, they fail to bark at the front door even when there is someone there (a couple weeks ago, I ordered pizza delivery for dinner, and they did not even notice when the delivery person arrived to drop the pizza off). This is a Type 2 error. They do not detect an effect even though it exists in reality.

when the true population mean is 3.0 is less than 0.05. Similarly, if alpha is set to 0.01 and we reject the null hypothesis, then this means that $p < 0.01$ and the probability of getting a sample mean of 3.3 with a true population mean of 3.0 is less than 0.01. *P*-values are usually reported when we write about the results of hypothesis tests. For example, I might write the following about the two-tailed test that I just conducted: "A *t*-test comparing the mean GPA of a random sample of 50 study abroad returnees, 3.3, to the hypothesized mean GPA, 3.0, indicated that the sample mean GPA was significantly different from the hypothesized mean GPA ($t = 5.00$, df = 49, $p < 0.05$)." Notice that in this example, I also provided readers with my calculated *t*-statistic and degrees of freedom in parentheses.

Confidence intervals. Alpha is also an important component for constructing confidence intervals around a sample mean. While the *t*-tests we have been working with until now are useful for exploring the difference between a specific sample mean and a hypothesized population mean, researchers often find it useful to estimate a range of values that likely contains the population mean. We define this range of values, called a **confidence interval**, using the sample mean, which is the center of the confidence interval, and the standard error of the mean, which determines how wide the confidence interval is. Confidence intervals are useful when we want to be more conservative in our guesses regarding the population mean (providing a range of values that likely includes the population mean rather than drawing from a single estimate of this mean derived from a sample) and more transparent about how much variation there is in our sample. An example will make this concept more concrete.

The formula[4] to calculate a confidence interval is as follows:

$$CI = \overline{X} \pm (t_\alpha)(s_{\overline{x}})$$

In this formula, $\overline{X}$ is the sample mean, the center of the confidence interval. t_α is the critical *t*-value associated with a particular alpha level, which denotes how confident we want to be that the confidence interval contains the population mean. The two most common alpha values for constructing confidence intervals in education research are 0.05 and 0.01. If we want to be confident that the confidence interval contains the unknown population mean 95% of the time, we set alpha to 0.05 (if we assume a large sample size, we know from the *t*-table in Appendix A that $t_\alpha = 1.96$—see the last row of this table. Note that, by default, confidence intervals are two-tailed, as values are assumed to fall on either side of the mean). If we want to be 99% confident that the confidence interval contains the unknown population mean, we set alpha to 0.01 (from the *t*-table in Appendix A, we know that for a large sample size $t_\alpha = 2.58$ for a 99% confidence interval). The last part of the confidence interval equation is $s_{\overline{x}}$, the estimated standard error. Notice that the larger the standard error, the wider the confidence interval will be. In other

---

[4] Technically, this formula is only appropriate when you do not know the population standard deviation and have to estimate the standard error using the sample standard deviation, an assumption we make throughout this chapter.

words, the more variation in the sampling distribution for the variable of interest, the less certain we can be about the range of values that contains the population mean.

We can construct a 95% confidence interval, for our sample mean GPA of 3.3 using this formula (remember that the estimated standard error of the sampling distribution was 0.06). In this case, our critical t-value (df = 49) for our two-tailed hypothesis test is 2.02 (t_α).

$$CI = \overline{X} \pm (t_\alpha)(s_{\overline{x}}) = 3.3 \pm (2.02)(0.06) = 3.3 \pm 0.12 = [3.18, 3.42]$$

In this example, the lower bound of the confidence interval is 3.18 and the upper bound is 3.42. In other words, we are 95% confident that, given a sample mean of 3.3 and an estimated standard error of 0.06, the population mean GPA for study abroad returnees falls between 3.18 and 3.42. Notice that these results are consistent with the two-tailed hypothesis test that I conducted earlier in this section. That is, this confidence interval does not contain a GPA of 3.0.

International educators can use information garnered from one-sample t-tests to support decision-making, and researchers can use this information to answer questions that advance knowledge in the field of international education. In this section, our running example has indicated that the mean GPA of study abroad returnees is likely significantly different from—and higher than—the hypothesized population mean GPA of 3.0. Remember that we hypothesized a mean GPA of 3.0 because this is the mean GPA for all students enrolled at this institution—both study abroad returnees and other students. Are study abroad returnees benefiting from grade inflation that happens when their grades are converted from international institutions? Or are there other explanations for this difference?

4.3 Other *t*-Tests

The basic process behind the **one-sample *t*-test** that we just conducted can be expanded to include other kinds of mean comparisons. For example, we might want to compare the mean GPAs of two populations of students, such as those who have applied to study abroad and those who have not. Similarly, we might want to compare the mean GPAs of a population of students the semester before and the semester after they participate in study abroad. These two scenarios describe the contexts wherein we would want to conduct a **two-samples *t*-test** and a **dependent samples *t*-test**, respectively.[5] The main difference between these two kinds of t-tests and the one-sample t-test that we just conducted is the formula for calculating the t-statistic. Once you have calculated the t-statistic, finding and comparing against the appropriate critical t-value in Appendix A follows essentially the same steps. Important to note here is that to even *consider* using these hypothesis tests

---

[5] These are not the only additional mean-comparison hypothesis tests, but they are likely the most useful for readers of this book.

to make causal claims, such as "study abroad *causes* a student's GPA to be higher", the samples under consideration must be random. In this scenario, you would have to randomly assign students to study abroad. Since this is rarely the case in international education, the results of these tests should be taken as simply descriptive—while study abroad applicants may have a significantly higher mean GPA compared to non-applicants, this does not mean that study abroad has anything to do with students' academic achievement. Similarly, while study abroad participants may have a higher mean GPA after study abroad, this does not mean that study abroad itself caused this increase in academic achievement.

4.3.1 Two-Samples *t*-tests

The formula for a two-samples *t*-test, which is used to test hypotheses about the differences in means of two groups, is as follows:

$$t = \frac{(\overline{X}_1 - \overline{X}_2) - 0}{s_{\overline{x}1 - \overline{x}2}}$$

In this equation, the numerator is the difference between the means of the two samples, such as the mean GPA of study abroad applicants ($\overline{X}_1$) and the mean GPA of non-applicants ($\overline{X}_2$) minus the hypothesized population difference in means, which is usually 0 (we could hypothesize other differences in means if we wanted to, but this is not typical). The denominator of this equation is the estimated standard error of the difference between these two means rather than the estimated standard error of a single mean, which accounts for the fact that there are two samples instead of one. When sample sizes are roughly equal (e.g., if I have one random sample of 50 study abroad applicants and another random sample of 50 non-applicants), the formula for estimating the standard error of the difference between two sample means is the following[6]:

$$s_{\overline{x}1-\overline{x}2} = \sqrt{s_{\overline{x}1}^2 + s_{\overline{x}2}^2}$$

---

[6] The formula for the estimated standard error of the difference between two independent sample means when sample sizes are not equal is considerably more complex:

$$s_{\overline{x}1-\overline{x}2} = \sqrt{\frac{SS_1 + SS_2}{n_1 + n_2 - 2}\left(\frac{1}{n_1} + \frac{1}{n_2}\right)}$$

This formula also assumes that population standard deviations are equal. Here, SS_1 and SS_2 refer to the sum of squared deviations for each sample, while n_1 and n_2 are the two sample sizes. Essentially, what this formula does is weight the standard error to account for the fact that the sample sizes are unequal. Differences in sample sizes are especially problematic when one sample's standard error is dramatically different from the standard error of the other sample (a violation of the homogeneity of variance assumption). This can lead to very misleading results. When you calculate a two-samples *t*-test using a statistical software program, this adjustment for sample size, if needed, is made for you automatically. It also is possible to compute a standard error that allows for unequal variances between samples.

4.3 Other t-Tests

Table 4.1 GPA statistics for study abroad applicants and non-applicants

	Study abroad applicants	Non-applicants
Mean GPA	3.30	2.60
Standard deviation	0.41	0.54
Standard error	0.06	0.08
N	50	50

Note Data are invented for illustrative purposes

Here, $s_{\bar{x}1}$ is the estimated standard error corresponding to the first sample (study abroad applicants) and $s_{\bar{x}2}$ is the estimated standard error corresponding to the second sample (non-applicants).

Table 4.1 contains information about two random samples of 50 students each, one of study abroad applicants and another of non-applicants. We will use information from these two samples to evaluate the following null and alternative hypotheses:

H_0: The difference in mean GPAs for study abroad applicants and non-applicants is zero.

H_A: The difference in mean GPAs for study abroad applicants and non-applicants is not zero.

These two hypotheses apply to the populations of study abroad applicants and non-applicants. Our two samples will help us make inferences about these populations.

We start by estimating the standard error of the difference between our sample means:

$$s_{\bar{x}1-\bar{x}2} = \sqrt{s_{\bar{x}1}^2 + s_{\bar{x}2}^2} = \sqrt{0.06^2 + 0.08^2} = \sqrt{0.0036 + 0.0064} = \sqrt{0.01} = 0.1$$

Next, we use this estimated standard error in the formula for t (notice that I drop the -0 from the numerator since mathematically it does not matter):

$$t = \frac{\bar{X}_1 - \bar{X}_2}{s_{\bar{x}1-\bar{x}2}} = \frac{3.3 - 2.6}{0.1} = \frac{0.7}{0.1} = 7.00$$

Once we calculate this t-statistic, we find the critical t-value in the table in Appendix A. To determine degrees of freedom, we first add the sizes of our two samples together ($50 + 50 = 100$) and then, to account for the fact that we have means from two samples rather than one, we subtract two from this overall sample size ($100 - 2 = 98$). Because our sample size is quite large, many at this point would feel safe in referring to the last row of the critical t-value table in Appendix A. If we want to be abundantly conservative, though, we have the option of again going to the next-lower row, which is df = 60. Note that our null hypothesis is a two-tailed one—we hypothesize that the population means are different from one another, not that one is higher or lower than the other. Assuming $\alpha = 0.05$ and this more conservative approach, we find a critical t-value of 2.00. (This would be 1.96

if we were happy to assume that our sample size was large enough to accommodate a normally-distributed test.) Since our calculated t-value, 7.00 is higher than both 2.00 and 1.96, we reject the null hypothesis and conclude that the difference in mean GPAs of study abroad applicants and non-applicants is significantly different from zero ($t = 7.00$, $p < 0.05$).

4.3.2 Dependent Samples *t*-test

A dependent samples t-test is very similar to a two-samples t-test in that we are interested in whether the difference between two sample means is discernible from a hypothesized difference of 0. The distinction between the two tests is that while a two-samples t-test compares two mutually exclusive groups (a student is either a study abroad applicant or not), a dependent samples t-test compares means for the same group at two different points in time. For this t-test, we have to account for the fact that the units in our sample are the same at Time 1 and Time 2, and as such, their data at each time are likely related (e.g., Student 1 at Time 1 has many of the same characteristics as Student 1 at Time 2 since this student is the same individual each time).

Here, we continue with an example based on study abroad participants but simplify our sample to include only five students for the sake of illustration. The means that we will compare correspond to the GPAs of study abroad participants before and after studying abroad. The data for this example are summarized in Table 4.2. As this table shows, three students (Student 1, Student 2, and Student 3) reported higher GPAs after studying abroad compared to before (as indicated in the next-to-last column of the table, which subtracts the student's GPA at Time 2 from their GPA at Time 1). One student (Student 4) did not experience a change in GPA over time, and another student (Student 5) experienced a decrease in GPA. On average, the mean GPA for this group of students increased from 3.04 to 3.24 between Time 1 and Time 2. The hypotheses that we will explore are the following (notice that this is a two-tailed test):

H_0: The difference in mean GPAs for study abroad participants before and after study abroad is zero.

H_A: The difference in mean GPAs for study abroad participants before and after study abroad is not zero.

Again, these hypotheses refer to differences in the population means of this group of students before and after studying abroad. We will use information from our sample to make inferences about the population.

The formula for a dependent samples t-test is as follows:

$$t = \frac{(\overline{X} - \overline{Y}) - 0}{s_{\overline{D}}}$$

In this equation, $\overline{X}$ is the mean GPA from our sample at Time 1 and $\overline{Y}$ is the mean GPA from our sample at Time 2. In the numerator, notice that we are again

4.3 Other t-Tests

Table 4.2 Study abroad participants' GPAs before (Time 1) and after (Time 2) studying abroad

	Time 1	Time 2	Difference (Time 2 − Time 1)
Student 1	3.30	3.50	0.20
Student 2	3.10	3.60	0.50
Student 3	2.70	3.10	0.40
Student 4	2.10	2.10	0.00
Student 5	4.00	3.90	-0.10
Mean	$\overline{X} = 3.04$	$\overline{Y} = 3.24$	
SD			0.26

Note Data are invented for illustrative purposes

hypothesizing that this difference in means is 0. $s_{\overline{D}}$ is the estimated standard error of the difference between dependent sample means. This estimated standard error is calculated exactly like the estimated standard error in a one-sample t-test, but, through a bit of numerical trickery, uses the standard deviation of the *differences* in values between Time 1 and Time 2 (s_D) (the last column of Table 4.2) rather than the standard deviation of actual GPA values. In accounting for the *differences* between Time 1 and Time 2, a dependent samples t-tests accounts for the fact that the students in our sample at Time 1 are the same students at Time 2. The formula for the estimated standard error of the difference in means is as follows:

$$s_{\overline{D}} = \frac{s_D}{\sqrt{n}}$$

I use the data in Table 4.2 to illustrate these calculations. First, I estimate the standard error of the differences between means using the information in the last column of Table 4.2:

$$s_{\overline{D}} = \frac{s_D}{\sqrt{n}} = \frac{0.26}{\sqrt{5}} = 0.12$$

Next, I use this number in the denominator of the formula for t. In the numerator, I subtract the sample mean for Time 1 from the sample mean for Time 2:

$$t = \frac{\overline{X} - \overline{Y}}{s_{\overline{D}}} = \frac{3.24 - 3.04}{0.12} = \frac{0.2}{0.12} = 1.67$$

To determine the critical value for this t-test, I use the number of paired groups ($N = 5$) in my sample rather than the number of total observations ($N = 10$) to calculate degrees of freedom. Subtracting 1 ($5 - 1$), I arrive at 4 degrees of freedom. If I set $\alpha = .05$, my critical t-value for a two-tailed t-test is 2.78. Since my calculated t-value (1.67) is less than 2.78, I fail to reject the null hypothesis and conclude that there is no evidence that the difference between students' mean

GPAs before and after study abroad is significantly different from zero ($t = 1.67$, $df = 4$, $p > 0.05$). Note that we make this decision even though the students in our sample have a higher mean GPA after studying abroad compared to before. What our hypothesis test tells us is that we do not have the evidence we need to determine that this difference is meaningful in the statistical sense.

4.4 Example from the Literature

An example of a research study that uses t-tests as a primary analytic technique is Yang et al. (2020). This study examines survey data from both first-year students ($N = 728$) attending an international branch campus institution in Malaysia and faculty who worked at the same institution ($N = 124$). The purpose of the survey was to better understand student and faculty expectations regarding first-year course content, faculty responsiveness, faculty and staff concern for the student, faculty assessment feedback, student preparation for assessment, personal interaction between faculty and students, and library support. Respondents were asked to rate statements about these constructs on a scale of from 1 to 5. Because students transition from their home country's education system to the branch campus institution, which is heavily influenced by a foreign institution, the authors hypothesized that first-year students and faculty would differ on their expectations about what students' first-year university experiences should be like. The authors conducted a series of independent samples t-tests (another way to refer to two-samples t-tests) to compare survey responses from students and faculty. Table 4.3 repro-

Table 4.3 Comparison of student and teacher responses for each construct (adapted from Yang et al., 2020)

Construct	Student mean (SE)	Faculty mean (SE)	p-values
Course content	4.302 (0.019)	3.815 (0.056)	0.000
Responsiveness	4.006 (0.023)	4.229 (0.041)	0.000
Feedback	4.148 (0.026)	4.181 (0.048)	0.536
Preparation for assessment (material support)	4.246 (0.025)	3.618 (0.064)	0.000
Preparation for assessment (effort required)	4.101 (0.021)	4.051 (0.042)	0.280
Personal interaction	4.215 (0.024)	3.720 (0.054)	0.000
Concern for student (faculty level)	3.983 (0.025)	3.944 (0.054)	0.534
Concern for student (staff level)	3.835 (0.029)	3.984 (0.069)	0.049
Library support (software)	4.437 (0.023)	4.263 (0.054)	0.003
Library support (hardware)	4.215 (0.028)	4.411 (0.052)	0.001

duces, in part, these *t*-tests (this information was taken from Table 4 on page 363 in Yang et al. [2020]).

As the small *p*-values in the last column of Table 4.3 indicate, results suggest significant differences in student and faculty ratings regarding expectations for several constructs. Specifically, students and faculty rated items corresponding to course content, responsiveness, preparation for assessment in terms of material support, personal interaction, staff concern for students, and library support (both software and hardware) differently. For example, regarding library support, students seemed to expect more from the library regarding software when compared to faculty. Students' mean rating for library software support was 4.437 compared to 4.263 for faculty ($p < 0.01$). In contrast, the average faculty rating for library hardware support was significantly higher compared to students' average rating (a mean of 4.411 compared to a mean of 4.215, respectively, $p < 0.001$). Notice that Table 4.3 contains all the information you would need to arrive at the *p*-values that Yang et al. (2020) report yourself—that is, this table provides the sample means for both groups and the sample standard errors for both groups. This way of reporting information is common in statistics so that readers can potentially compare your results to results from their own data. Yang et al. (2020) conclude that the information that their study provides can be helpful to leaders at international branch campus institutions in making changes to better align student and faculty expectations. In turn, more aligned expectations are expected to improve the learning experience for everyone involved.

4.5 Practice Problems

1. An administrator makes the claim that international students at my institution have an average GPA of 2.5. To test if this hypothesis is correct, I draw a random sample of 10 students and find that their average GPA is 3.09 with a standard deviation of 0.57.
    a. What are the null and alternative hypotheses (two-tailed) for a comparison between this sample GPA and the population average?
    b. What is the estimated standard error of the mean for this distribution?
    c. What is the calculated *t* needed to test the null hypothesis in (a)?
    d. What is the critical *t* value needed to test the null hypothesis in (a) if $\alpha = 0.05$?
    e. Do you reject or fail to reject the null hypothesis?
    f. Construct a 95% confidence interval for the sample mean.
2. The table below provides information about the number of globalized courses two groups of community college students took during their degree programs. On average, students in arts and sciences degree programs took more globalized courses ($\overline{X}_1 = 2.8$) compared to those in applied degree programs ($\overline{X}_2 = 1.3$).

	Arts and sciences degree programs	Applied degree programs
Average number of globalized courses	2.8	1.3
N	100	100
SE	0.3	0.8

Note Data in this table are invented for illustrative purposes

    a. What are the null and alternative hypotheses (two-tailed) for a comparison between the mean number of globalized courses taken by these two groups of students?
    b. What is the standard error of the difference between the two sample means?
    c. What is the calculated t needed to test the null hypothesis in (a)?
    d. What is the critical t value needed to test the null hypothesis in (a) if $\alpha = 0.01$?
    e. Do you reject or fail to reject the null hypothesis?

3. Sample Dataset #2 contains information about study abroad participation (study abroad) at 161 U.S. liberal arts institutions in the 2016–17 academic year. Using the statistical software of your choice, conduct t-tests that examine the following null hypotheses. Write your conclusions out in words.
    a. The mean study abroad participation at liberal arts institutions is equal to 200.
    b. The mean study abroad participation at liberal arts institutions is equal to 300.
    c. The difference between mean study abroad participation at liberal arts institutions in the highest SAT category (SAT700plus) and the mean study abroad participation at all other liberal arts institutions is zero.
    d. The difference between mean study abroad participation at liberal arts institutions with no reported SAT information (SAT_none) and the mean study abroad participation at all other liberal arts institutions is zero.

Recommended Reading

A Deeper Dive

Urdan, T. C. (2017a). Standard errors. In *Statistics in plain English* (pp. 57–72). Routledge.
Urdan, T. C. (2017b). Statistical significance, effect size, and confidence intervals. In *Statistics in plain English* (pp. 73–92). Routledge.
Urdan, T. C. (2017c). *t* Tests. In *Statistics in plain English* (pp. 73–112). Routledge.

Wheelan, C. (2013a). Basic probability: Don't buy the extended warranty on your $99 printer. In *Naked statistics: Stripping the dread from data* (pp. 68–89). Norton.

Wheelan, C. (2013b). The Monty Hall problem. In *Naked statistics: Stripping the dread from data* (pp. 90–94). Norton.

Wheelan, C. (2013c). Problems with probability: How overconfident math geeks nearly destroyed the global financial system. In *Naked statistics: Stripping the dread from data* (pp.95–109). Norton.

Wheelan, C. (2013d). The central limit theorem. In *Naked statistics: Stripping the dread from data* (pp. 127–142). Norton.

Additional Examples

Cartwright, C., Stevens, M., & Schneider, K. (2021). Constructing the learning outcomes with intercultural assessment: A 3-year study of a graduate study abroad and glocal experience programs. *Frontiers: The Interdisciplinary Journal of Study Abroad, 33*(1), 82–105.

Echcharfy, M. (2020). Intercultural learning in Moroccan higher education: A comparison between teachers' perceptions and students' expectations. *International Journal of ResEarch in English Education, 5*(1), 19–35.

Yang, L., Borrowman, L., Tan, M. Y., & New, J. Y. (2020). Expectations in transition: Students' and teachers' expectations of university in an international branch campus. *Journal of Studies in International Education, 24*(3), 352–370.

One-Way ANOVA and the Chi-Square Test of Independence

Chapter 4 introduced hypothesis testing, our first step into inferential statistics, which allows researchers to take data from samples and generalize about an entire population. This chapter presented the sampling distribution, along with its associated properties, the expected value of the mean and estimated standard error. The Central Limit Theorem helped us to make use of the sampling distribution to test hypotheses about variables even when they are not normally distributed. The remainder of Chap. 4 illustrated hypothesis testing with several different kinds of t-tests, namely one-sample t-tests, two-sample t-tests, and dependent sample t-tests. Each of these hypothesis tests takes the same general form: it is the ratio of a comparison of means (or difference in means) to the amount of variability in the measure of interest (the estimated standard error). That is, to calculate a t-test, we divide a mean comparison by the standard error, which is calculated in different ways depending on what kind of hypothesis test we need. We will see this pattern over and over throughout the remainder of this book—to conduct a hypothesis test and make decisions about statistical significance, we account for our comparison of interest in the numerator and the variability in our data in the denominator.

This chapter introduces two additional types of hypothesis tests. First, a **one-way analysis of variance (ANOVA)** expands on the two-samples t-test from the previous chapter and considers that we may want to compare the means among more than two groups. In this illustration, we will build on our running example that compared the mean GPAs of study abroad applicants and non-applicants to include a third group of students that applied to participate in a virtual exchange program. In this section, we will consider whether the mean GPAs of students significantly differ across the three groups (study abroad, virtual exchange, and neither option). Second, a **chi-square test of independence** considers that, while we might be interested in comparing multiple groups to one another, we may not always be interested in comparing these groups in terms of continuous variables, like GPA. In brief, a chi-square test of independence is a kind of nonparametric statistic (more on nonparametric statistics later) that is used to compare two categorical variables. A chi-square test would be appropriate, for example, in a situation where we wanted to explore the kinds of international learning opportunities (a categorical variable) that students identifying with different racial/ethnic groups (another categorical variable) choose to undertake.

5.1 One-Way Analysis of Variance (ANOVA)

To illustrate how one-way ANOVA works, we expand on our example from the previous chapter, which compared the mean GPAs of study abroad applicants and non-applicants, to include a third group, students who applied to participate in a virtual exchange program. Virtual exchange programs have the potential to provide students with many of the benefits of study abroad without much of the extra costs. At the same time, these exchange programs arguably lack many important elements of experiential learning available to students through study abroad (e.g., the opportunity to live with a host family). Consequently, it is important for international educators to consider key differences in the characteristics of students who apply to these opportunities to address questions of equitable access to international learning. However, similar to the t-tests in the previous chapter, the results of the examples in this chapter should not be interpreted as causal. GPA does not cause a student to select one particular learning opportunity over another—rather, our analysis will describe general trends in the characteristics of different student groups. For illustrative purposes, we will use the data in Table 5.1 to show how a one-way ANOVA works and to test whether the mean GPAs among these three groups of students (study abroad, virtual exchange, neither) are significantly different from one another.

Notice that Table 5.1 contains GPA data for five students in each program context—study abroad, virtual exchange, and neither option. At the bottom of the table, the last row calculates the mean GPA for each group. We see that these means pattern as we might expect, especially given what we know about students who are able to afford to participate in study abroad (e.g., that they come from relatively well-off backgrounds and likely have access to resources that would

Table 5.1 GPA data for study abroad applicants, virtual exchange applicants, and students applying to neither opportunity

	Study abroad	Virtual exchange	Neither
1	3.4	3.2	2.5
2	2.8	2.1	2.1
3	2.7	2.3	1.3
4	3.9	2.8	3.5
5	2.9	2.3	2.7
Mean	3.14	2.54	2.42

Note Data in this table are invented for illustrative purposes

enhance their academic performance even before studying abroad), the mean GPA for study abroad applicants is the highest ($\overline{X}_{SA} = 3.14$), followed by the mean GPA for virtual exchange applicants ($\overline{X}_{VE} = 2.54$), and, lastly, the mean GPA for neither program ($\overline{X}_{NP} = 2.42$). The between-group comparison is what is of interest to us—we want to know if these three groups are meaningfully different from one another when it comes to academic achievement, as measured using GPA. We can visually compare the mean GPAs of our three groups and see that study abroad applicants have the highest mean GPA, virtual exchange applicants have the second highest, and the neither group has the lowest—but, are these differences meaningful once we account for within-group variation in GPA (also called the error)? If GPA varies widely within each of the three groups, these differences in means could simply be due to idiosyncratic characteristics of the samples. In this sense, conducting a one-way ANOVA is no different from the *t*-tests in Chap. 4—we calculate the ratio of our comparison of interest to the error within groups. Calculating a one-way ANOVA can be usefully divided into three distinct steps.

5.1.1 Step One: Between-Group Variation

The first step in a one-way ANOVA analysis is to calculate the variation that we are interested in knowing about—the differences between group means. To do this, we calculate what is called the **mean square between** (MS_b) (you can think of this as being shorthand for the mean square between error—the difference between groups). Just like in any other situation where you would need to find a mean, the first step in finding the MS_b is finding the *total* amount of between-group variation, referred to formally as the **sum of squares between groups** (SS_b). SS_b is calculated by first subtracting the **grand mean** ($\overline{X}_T$), which refers to the mean for all groups combined (the mean GPA for all the students in our sample, regardless of their intercultural learning experience choice), from each group mean ($\overline{X}$) and squaring these differences. Next, we account for the number of observations in each group and finally sum the squared differences. You might recognize this process as being very similar to how we calculated variance in Chap. 3, where we calculated the sum of the squared differences between individual observations in a dataset and the mean of these observations.

Table 5.2 illustrates how to calculate SS_b for our example data. The first column of this table calculates the first step in this process: subtracting the grand mean from each group mean. In this case, the grand mean for our data is 2.7 (40.5/15). For example, in the case of study abroad participants (the first row), we calculate $3.14 - 2.7$ to arrive at 0.44. In the second step, we square these differences (the square of 0.44 is 0.1936). The third step involves a correction for the sample size of each group, which helps us account for the fact that we may be comparing samples of very different sizes (e.g., consider a more realistic scenario where only a few students participate in study abroad or virtual exchange while the vast majority of students at a given institution of higher education participates in neither of these experiences). To do this, we multiply the squared differences from Step 2 by the

Table 5.2 Sum of squares between groups calculation for example data

Group	Step 1: Subtract the grand mean from the group mean	Step 2: Square the differences	Step 3: Account for sample size (n) of each group
	$(\overline{X} - \overline{X}_T)$	$(\overline{X} - \overline{X}_T)^2$	$n(\overline{X} - \overline{X}_T)^2$
Study abroad	3.14–2.7 = 0.44	$0.44^2 = 0.1936$	$0.1936 * 5 = 0.968$
Virtual exchange	2.54–2.7 = −0.16	$-0.16^2 = 0.0256$	$0.0256 * 5 = 0.128$
Non-participation	2.42–2.7 = −0.28	$-0.28^2 = 0.0784$	$0.0784 * 5 = 0.392$
	Step 4: Sum squares between groups	$\sum_{i=1}^{K} \left[n(\overline{X} - \overline{X}_T)^2 \right]$	$0.968 + 0.128 + 0.392 = 1.488$

Note Data in this table are invented for illustrative purposes

number of observations in each group (in this case, n = 5 for all three groups, so 0.1936*5 = 0.968). Finally, in Step 4 (represented in the row at the bottom of Table 5.2), we add these squared differences together, arriving at the sum of squared differences, or the sum of squares, between groups.

To convert the sum of squares between groups (SS_b) that we just calculated into the mean of squared between differences (MS_b) that we need for our hypothesis test, we divide SS_b by the appropriate number of degrees of freedom, just like we have done in previous calculations. In this case, we subtract 1 from the total number of *groups* that we are comparing, represented with K in standard statistical notation. Thus, MS_b is formally calculated as follows (where $K - 1$ represents degrees of freedom):

$$MS_b = \frac{SS_b}{K - 1}$$

In our example, we divide 1.488 (SS_b) by 2 (since we are comparing three groups, $3 - 1 = 2$) to arrive at 0.744.

5.1.2 Step Two: Within-Group Variation

The second step in a one-way ANOVA analysis is to calculate the variation within each group. If this variation is substantial, then differences in group means that appear large may not be all that meaningful in the statistical sense. That is, while the measure of between-group variation that we just calculated is the difference in group means that we are interested in (i.e., the differences in group mean GPAs in Table 5.1), we have to account for the variation within groups to know if this difference among means is statistically significant. Within-group variation is known as the *error* in an ANOVA analysis and is the denominator in our hypothesis test calculation. To find the average amount of variation *within* groups, we calculate what is called the **mean square within** (MS_w) (also referred to as the mean square error in some statistics books). A calculation of MS_w starts with the **sum of squares**

5.1 One-Way Analysis of Variance (ANOVA)

within groups (SS_w), which is the within-group equivalent of the sum of squares *between* groups from Step 1.

To calculate SS_w, we first subtract each group mean from the individual observation values within that group. Then, we square these deviations from the mean (this process is equivalent to what we did to calculate variance in Chap. 3). In a third step, we sum these squared values for each group and, as a fourth step, add the summed squares for all groups together. Table 5.3 illustrates these calculations for our running example. In the first step (represented in the first column for each group), we subtract the group mean from each individual value within the group ($X - \overline{X}$). Notice that in the first column, we subtract 3.14, the mean GPA for study abroad participants, from each individual study abroad applicant's GPA. As a second step, we square these values (($X - \overline{X})^2$) and then, in step 3, we sum these within group squares, arriving at a sum of squares for each group—three, in our

Table 5.3 Sum of squares within groups calculation for example data

	Study abroad ($\overline{X} = 3.14$)		Virtual exchange ($\overline{X} = 2.54$)		Non-participation ($\overline{X} = 2.42$)	
	Step 1: Subtract the group mean from each individual value	Step 2: Square the differences	Step 1: Subtract the group mean from each individual value	Step 2: Square the differences	Step 1: Subtract the group mean from each individual value	Step 2: Square the differences
	$X - \overline{X}$	$(X - \overline{X})^2$	$X - \overline{X}$	$(X - \overline{X})^2$	$X - \overline{X}$	$(X - \overline{X})^2$
1	3.4 − 3.14 = 0.26	0.068	3.2 − 2.54 = 0.66	0.436	2.5 − 2.42 = 0.08	0.006
2	2.8 − 3.14 = −0.34	0.116	2.1 − 2.54 = −0.44	0.194	2.1 − 2.42 = −0.32	0.102
3	2.7 − 3.14 = −0.44	0.194	2.3 − 2.54 = −0.24	0.058	1.3 − 2.42 = −1.12	1.254
4	3.9 − 3.14 = 0.76	0.578	2.8 − 2.54 = 0.26	0.068	3.5 − 2.42 = 1.08	1.166
5	2.9 − 3.14 = −0.24	0.058	2.3 − 2.54 = −0.24	0.058	2.7 − 2.42 = 0.28	0.078
Step 3: Sum the squared differences for each group	$\sum(X - \overline{X})^2 =$	1.012	$\sum(X - \overline{X})^2 =$	0.812	$\sum(X - \overline{X})^2 =$	2.608
Step 4: Sum of sums of squared differences	1.012 + 0.812 + 2.608 = 4.432					

Note Data in this table are invented for illustrative purposes

case (this last step is found in the last row of Table 5.3) ($\sum(X - \overline{X})^2$). The fourth and final step for calculating SS_w is to add these summed values for all groups together: $1.012 + 0.812 + 2.608 = 4.432$ (see the last row of Table 5.3).

Once we calculate SS_w, we again divide by the appropriate number of degrees of freedom to arrive at MS_w, as follows:

$$MS_w = \frac{SS_w}{N - K}$$

This formula shows us that degrees of freedom for MS_w is $N - K$, thus accounting for both the number of observations in our dataset (N) and the number of groups (K) that we are comparing. In our case, since we have 15 individuals in the dataset (five per group) and three groups, we use 12 degrees of freedom ($15 - 3 = 12$). MS_w, then, is $4.432/12$, or approximately 0.369.

5.1.3 Step Three: The F-distribution

We now have all the pieces we need to conduct a hypothesis test exploring the null hypothesis that the mean GPAs of our three groups, study abroad, virtual exchange, and neither option, are equal. Keep in mind that this hypothesis is at the population level, and we are using our three samples to make inferences. We have a measure of the mean differences between the groups we want to compare ($MS_b = 0.744$) and a measure of the mean variation within the groups ($MS_w = 0.369$). We can take the ratio of MS_b and MS_w to calculate our test statistic, arriving at 2.02 ($0.744/0.369 = 2.02$). If this were a t-test, we would turn to the table of critical t-values in Appendix A, compare our calculated statistic to the critical value for our alpha level of choice (e.g., $\alpha = 0.05$), and make a decision about rejecting or failing to reject the null hypothesis. However, notice one key difference between the test statistic we calculate in an ANOVA and the test statistic we calculate in a t-test: The test statistic in an ANOVA is calculated using sums of *squared* deviations. For this reason, we cannot use the t-distribution to conduct hypothesis tests in an ANOVA, we have to use the square of the t-distribution, which is called the F-distribution, illustrated in Fig. 5.1. Like the t-distribution, the F-distribution takes different shapes depending on the degrees of freedom in a given calculation, which helps researchers account for both sample size (recall Fig. 4.4) and the number of groups in a comparison (notice that in Fig. 5.1, there are two degrees of freedom for each instance of the F distribution—one for the numerator and another for the denominator—degrees of freedom enter into the calculation of the F-distribution, which results in the different shapes you observe in Fig. 5.1). In our running example, we had two degrees of freedom in the numerator ($K - 1$) and twelve degrees of freedom in the denominator ($N - K$).

The test statistic that we calculate when we divide MS_b by MS_w in an ANOVA is called an F-statistic, and is formally represented as follows in equation form (remember we already calculated this test statistic for our running example in an informal sense, arriving at 2.02):

5.1 One-Way Analysis of Variance (ANOVA)

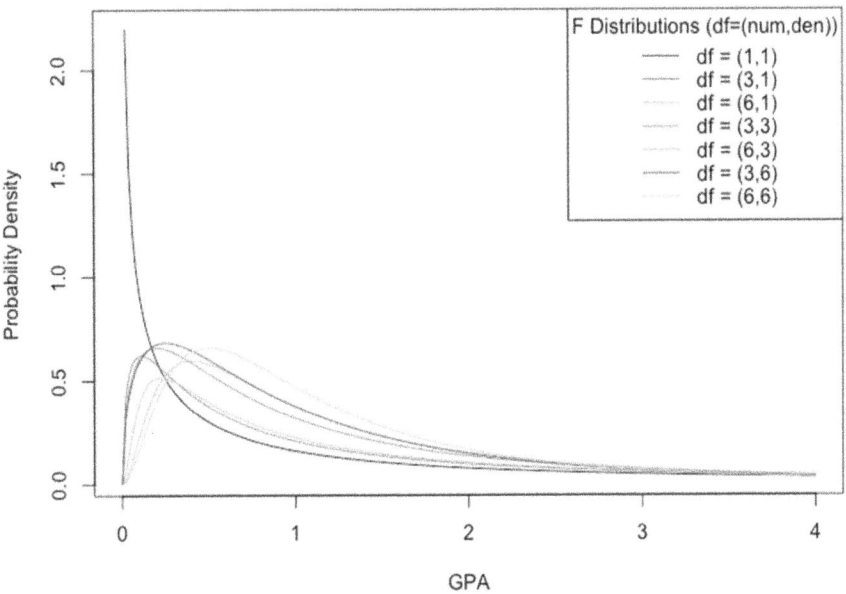

Fig. 5.1 The F-distribution at various degrees of freedom

$$F = \frac{MS_b}{MS_w}$$

Critical values for the F-distribution are found in Appendix B. Notice that in this table, critical values are found by accounting for the degrees of freedom in the numerator (across the top of the table) and the degrees of freedom in the denominator (down the side of the table), which makes this table very large. To simplify this table a bit, all critical values in this table correspond to $\alpha = 0.05$. However, in other statistics textbooks, you will find more complex tables that contain other alpha values (e.g., $\alpha = 0.01$). For our purposes, we find that the critical value corresponding to $df = 2$ in the numerator and $df = 12$ in the denominator is 3.885. Since our calculated F-value (2.02) is smaller than this critical value, we fail to reject the null hypothesis and conclude that we do not have sufficient evidence to assert that the group mean GPAs among students who apply to study abroad, to participate in virtual exchange, or to neither opportunity are different from one another.[1]

---

[1] If we had found a significant difference in mean GPA among the three groups in this example, a logical next question would be to ask what groups account for this significant difference. That is, is the significant difference between students who study abroad and students who participate in virtual exchange? Or perhaps a significant difference is found comparing all three groups to each other. A variety of tests called *post-hoc tests* are available to researchers to explore these differences. The appropriate post-hoc test for a given ANOVA analysis depends on many factors,

5.1.4 Example from the Literature

An example of a study that uses one-way ANOVA in its analysis is Strange and Gibson (2017). In this study, the authors were interested in whether transformative learning differed by the length of a student's study abroad program. They measured the amount of transformative learning that a student experienced during study abroad using an instrument that asked students to agree or disagree with twelve statements, such as "I had an experience that caused me to question the way I normally act." Students' responses to these statements were then aggregated to produce a single "transformative learning" measure for each student that ranged in value from 0 to 12. The authors then calculated mean transformative learning values for four groups of study abroad students according to the length of their study abroad program: short (0–18 days), medium (19–35 days), long (36–49 days), and extra-long (50 or more days) and conducted a one-way ANOVA to determine if there were differences in mean transformative learning when comparing these four groups. Descriptive information about students' transformative learning by program length are reproduced in Table 5.4. Results of Strange and Gibson's (2017) one-way ANOVA are reproduced in Table 5.5.

Notice in Table 5.5 that Strange and Gibson (2017) use 3 degrees of freedom (DF) when making between-group calculations because they compare four groups $(K - 1)$. Similarly, since they have 120 participants in their study, they calculate 116 degrees of freedom for their within-group calculations $(N - K)$. Dividing their

Table 5.4 Descriptive information regarding transformative learning and study abroad program length

Program Length	N	Mean	SD
Short	19	4.29	3.24
Medium	45	7.09	3.12
Long	36	7.26	2.79
Extra-long	25	7.92	3.19
Total	120	6.91	3.22

Reproduced from Table 2 in Strange and Gibson [2017]
Note Data in this table are invented for illustrative purposes

---

including whether the groups have similar or different sample sizes and whether the within-group variances for the groups are very similar to or very different from one another. Post-hoc tests are similar to *t*-tests in that they compare two group means, but these tests reduce the likelihood of Type 1 error (detecting a significant difference when it is not there) by accounting for the number of comparisons that are being made (otherwise, the more comparisons a researcher makes—and the more *t*-tests they conduct—the greater the likelihood that they will encounter a Type 1 error). While I do not recommend a specific post-hoc test in this chapter given the numerous tests available and their specificity to certain properties of the variables in the comparisons being made, these tests may be worth exploring if ANOVA is the end point of your analysis. However, if you are using ANOVA as a steppingstone to a more complex statistical analysis, such as one of the regression approaches introduced in Chaps. 7–9), your regression results will be a much more efficient way to explore the nature of a significant difference between groups.

5.2 Chi-Square Test of Independence

Table 5.5 One-way ANOVA results examining transformative learning and study abroad program length

	Sum of squares	DF	Mean square	F
Between groups	146.37	3	48.79	5.21
Within groups	1085.63	116	9.36	

Reproduced from Table 2 in Strange and Gibson [2017]
Note Data in this table are invented for illustrative purposes

between and within sums of squares calculations by these degrees of freedom produces the mean squares in the fourth column of Table 5.5. Dividing the mean square for the between group calculation by the mean square of the within group calculation produces the F-value in the last column of this table (5.21). Strange and Gibson (2017) provide the p-value associated with this calculated F-statistic in their article ($p = 0.002$) so we know that there is a significant difference in number of transformative learning experiences between at least two of their study abroad program length groups (we could even guess that it is the Short group that is different from the other groups, given the means provided in Table 5.4, but a post hoc test [see Footnote 1 in this chapter] would help us back up this assertion). Length of a student's abroad program, then, seems to matter when it comes to transformative learning experience. International educators might make use of this information when they design and offer study abroad programs.

While Strange and Gibson provide a p-value for their one-way ANOVA, we could also explore the significance of their calculated F-statistic using the F-table in Appendix B. The critical value for 3 degrees of freedom in the numerator and 116 degrees of freedom in the denominator falls somewhere between 2.6955 and 2.6771 (the values for 100 and 125 degrees of freedom in the denominator, respectively). Since 5.21 is larger than both these values, we can reject the null hypothesis that there is no difference in transformative learning among groups of students studying abroad for different amounts of time at $p < 0.05$. If I were writing this result up formally, I might write the following: "A one-way ANOVA suggested that the mean number of transformative learning experiences that students have during study abroad differed according to the amount of time they were abroad ($F(3,116) = 5.21, p < 0.05$)". Notice that in this sentence in the parentheses after F, I provide the degrees of freedom for the numerator and denominator for the F-test. In this way, if someone were to double check my work, they would be able to do so easily.

5.2 Chi-Square Test of Independence

An important feature of the hypothesis tests that we have conducted thus far is that we have been working with continuous variables, like a student's numerical GPA. However, often we want to compare groups on a categorical variable, such as racial/ethnic identification and type of international learning opportunity that a student chooses to participate in. That is, we may want to test the null hypothesis that racial/ethnic identification (one grouping variable) is not related to a student's selection of international learning opportunity (another grouping variable). In this case, it is unclear how we would even begin to calculate the mean values of a particular variable for each group. Here, we turn to a family of statistics referred to as **nonparametric**, which allow us to conduct a hypothesis test with two categorical variables.

In brief, the test statistics that we have introduced thus far rely on what are called **parameters**—a quantifiable characteristic of a population that is used in **parametric** statistics. Both the mean and the variance are examples of parameters. Parameters are used to calculate test statistics and construct sampling distributions (notice that we have been using these basic building blocks in our t-test and ANOVA calculations all along). Nonparametric statistics, on the other hand, do not rely on parameters and can be used when working with purely categorical data. Here, I introduce one kind of nonparametric statistic that is especially useful for international education researchers, the chi-square test of independence. As you might guess, this test explores the relationship between two categorical variables, such as racial/ethnic identity and international learning experience choice.

At the heart of a chi-square test of independence are the concepts of observed and expected frequencies. Observed frequencies are the values that you observe in a dataset. In the running example in this section, we will use the observed frequencies of participation in international education experiences by racial/ethnic identification in Table 5.6, which summarizes data for 200 students. To simplify our example, we include data from only three racial/ethnic identification groups: Black, white, and a group that includes all other students. However, it is important to acknowledge that this 'other' racial/ethnic group may be quite heterogenous in a particular context—including individuals who identify as Asian, Latinx, or Native American (and even more specific categories exist within these groups—e.g., Latinx may include students with Mexican or South American heritage, it may also include first-, second-, and third-generation immigrants, etc.). In practice, I recommend disaggregating data into as many groups as possible when conducting an analysis (keeping in mind that you need a certain number of individuals in each group for statistical tests to be valid—severe small samples is a condition sometimes referred to as *micronumerosity* in statistics), as these additional identities may be very important.

As Table 5.6 indicates, five Black students participated in study abroad, 12 participated in virtual exchange, and 19 did not participate in an international

Table 5.6 Observed frequency of participation in international learning experiences by racial/ethnic identification (N = 200)

	Study abroad	Virtual exchange	Non-participation
Black	5	12	19
White	15	36	53
Other	3	24	33

Note Data in this table are invented for illustrative purposes

5.2 Chi-Square Test of Independence

Table 5.7 Observed frequency of participation in international learning experiences by racial/ethnic identification with row and column sums (N = 200)

	Study abroad	Virtual exchange	No participation	Row sum
Black	5	12	19	36
White	15	36	53	104
Other	3	24	33	60
Column sum	23	72	105	

Note Data in this table are invented for illustrative purposes

learning experience. For White students, 15 studied abroad, 36 participated in virtual learning, and 53 did not participate in either opportunity. Finally, for students identifying with other racial/ethnic groups, three studied abroad, 24 participated in virtual exchange, and 33 did not participate in either opportunity. A chi-square test will tell us if this distribution differs from what we would obtain if we simply assigned students to international learning opportunities randomly—the expected frequency distribution.

5.2.1 Calculating Expected Frequencies

To compare observed frequencies, such as those in Table 5.6, to a distribution that would result from random chance, we have to calculate expected frequencies. We can then determine if the observed frequencies significantly differ from these expected frequencies.[2] To calculate expected frequencies, we first calculate the row and column sums from our observed distribution, as in Table 5.7.

In a second step, we use these row and column sums to calculate the expected value for each combination of groups in our sample—such as the expected number of Black students who would study abroad, the expected number of Black students who would participate in virtual exchange, and the expected number of Black students who would not participate in an international learning experience—if participation were randomly distributed. To calculate these expected values, we multiply the row sum by the column sum and then divide by the total number of units in the dataset (200 students, in this case) for each cell in the table ($\frac{(rowsum)*(columnsum)}{N}$). This calculation is illustrated in Table 5.8. Notice that in some cases, the expected values in Table 5.8 are not very different from the observed values in Table 5.6. For example, if international learning experiences were randomly distributed, we would expect around four Black students to study abroad. In reality, five Black students in our sample studied abroad. In other cases, expected and observed values are very different from one another. For example, Table 5.8 shows that we would expect seven students in the 'other' racial/ethnic

---

[2] Note that when expected frequencies are low, the chi-square test fails and is not a valid statistical approach.

Table 5.8 Calculation of expected frequency of participation in international learning experiences by racial/ethnic identification (N = 200)

	Study abroad	Virtual exchange	No participation	Row sum
Black	$\frac{36*23}{200} = 4.14$	$\frac{36*72}{200} = 12.96$	$\frac{36*105}{200} = 18.90$	36
White	$\frac{104*23}{200} = 11.96$	$\frac{104*72}{200} = 37.44$	$\frac{104*105}{200} = 54.60$	104
Other	$\frac{60*23}{200} = 6.90$	$\frac{60*72}{200} = 21.60$	$\frac{60*105}{200} = 31.50$	60
Column sum	23	72	105	

Note Data in this table are invented for illustrative purposes

group to study abroad. However, only three of these students participated in this international learning experience in our observed data.

5.2.2 Comparing Observed and Expected Frequencies

Once expected values have been calculated, our next task is to calculate a comparison of expected and observed frequencies for each combination of categories possible, that is, for each cell in Table 5.8. To calculate this comparison, we subtract the expected frequency from the observed frequency for each cell, square this value (as we have done before), and then divide our result by the expected frequency. You can think of this expected frequency in the denominator as the nonparametric substitute for the estimated error that appeared in the denominator in both *t*-tests and one-way ANOVA. In practice, the expected frequency in the denominator serves to scale the size of the difference between observed and expected frequencies: A one-unit difference between observed and expected frequencies could be quite dramatic for smaller values (say 2 versus 3), while a one-unit difference is less dramatic for larger values (say 1314 versus 1315). The general equation for this calculation is as follows:

$$\frac{(O - E)^2}{E}$$

where O is the observed value and E is the expected value. These calculations for our running example are illustrated in Table 5.9. Notice that the further away an observed value is from the expected value, the larger the number we calculate for that particular cell.

5.2.3 The Chi-Square (χ^2) Distribution

The remainder of a chi-square test proceeds in a way that is now likely very familiar. We need a critical value corresponding to the alpha level we have chosen for our analysis (here we will use $\alpha = 0.05$) and a calculated value based on

5.2 Chi-Square Test of Independence

Table 5.9 Calculation of χ^2 statistic

	Study abroad			Virtual exchange			No participation		
	Observed	Expected	$\frac{(O-E)^2}{E}$	Observed	Expected	$\frac{(O-E)^2}{E}$	Observed	Expected	$\frac{(O-E)^2}{E}$
Black	5	4.14	$\frac{(5-4.14)^2}{4.14} = 0.18$	12	12.96	$\frac{(12-12.96)^2}{12.96} = 0.07$	19	18.90	$\frac{(19-18.90)^2}{18.90} = 0.00$
White	15	11.96	$\frac{(15-11.96)^2}{11.96} = 0.77$	36	37.44	$\frac{(36-37.44)^2}{37.44} = 0.06$	53	54.60	$\frac{(53-54.60)^2}{54.60} = 0.05$
Other	3	6.90	$\frac{(3-6.90)^2}{6.90} = 2.20$	24	21.60	$\frac{(24-21.60)^2}{21.60} = 0.11$	33	31.50	$\frac{(33-31.50)^2}{31.50} = 0.07$

Note Data in this table are invented for illustrative purposes

our data. Similar to how we used the t-distribution to determine critical values for t-tests and the F-distribution to determine critical values for one-way ANOVA, we use the chi-square (χ^2) distribution for hypothesis testing in a chi-square test. The chi-square distribution is illustrated in Fig. 5.2 and critical values for the chi-square distribution are found in Appendix C. Notice that this distribution accounts for sample size in that it is adjusted for degrees of freedom, similar to both the t-distribution and the F-distribution. The appropriate number of degrees of freedom for a particular chi-square test accounts for the number of groups in both grouping variables in an analysis. Specifically, degrees of freedom in a chi-square test is calculated by multiplying the number of rows (R) minus one and the number of columns (C) minus one, $df = (R-1)(C-1)$. In our running example, we have 4 degrees of freedom ($df = (3-1)(3-1) = (2)(2) = 4$). The table in Appendix C indicates that at $\alpha = 0.05$ and $df = 4$, the critical chi-square value is 9.4877.

The last bit of information that we need to conduct our chi-square test is a calculated chi-square value. Calculated chi-square is simply the sum of all the values in the individual cells in Table 5.9. Formally, calculated chi-square is as follows:

$$\chi^2 = \sum \left(\frac{(O-E)^2}{E} \right)$$

In our running example, χ^2 is equal to 3.51 (0.18 + 0.77 + 2.20 + 0.07 + 0.06 + 0.11 + 0.00 + 0.05 + 0.07 = 3.51). Notice that this calculated value becomes large as the differences between individual pairs of observed and expected values become large. That is, the calculated chi-square is bigger as the differences between expected and observed values increase. Comparing our calculated chi-square value (3.51) to the critical chi-square value (9.4877) suggests that, in this

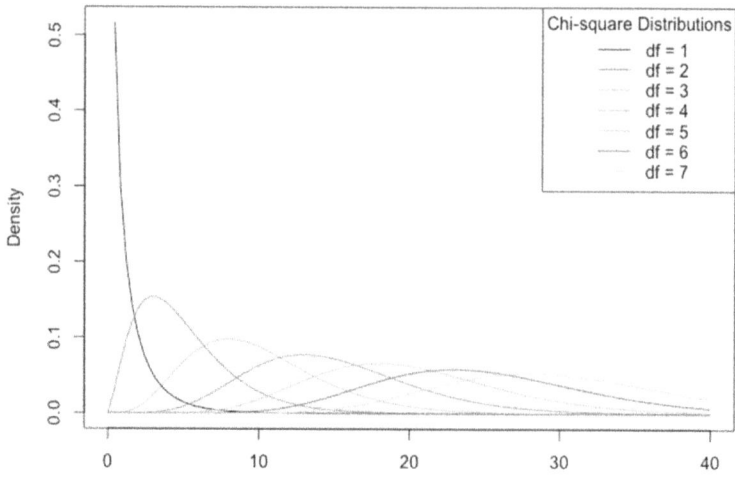

Fig. 5.2 Chi-square distribution at various degrees of freedom

5.2.4 Example from the Literature

Markham et al. (2017) used a series of chi-square tests to examine whether completing a class on second language learning theory changed the way that different groups of teachers-in-training approached how they planned to work with English language learners in their classrooms. Using data from a series of student homework assignments, researchers placed students in the study into four groups based on the extent to which their plans to incorporate second language learning theory into their classrooms changed over the course of the semester-long class: completely different (the elements of second language learning theory that students intended to incorporate into their classrooms were completely different when comparing the beginning to the end of the semester), somewhat different, mostly the same, and exactly the same. The researchers were especially interested in how the distribution of teachers-in-training into these four groups differed by their service status (pre-service or in-service) and whether they were considered domestic or international students at their higher education institution. The observed distribution corresponding to this latter comparison (domestic vs. international student status) is reproduced in Table 5.10. Notice that this table also provides the column-based percentage of students in each learning theory group, which helps illustrate how the distributions of students in each domestic/international group are different from one another. In this case, we see that a greater percentage of domestic students fell into the *somewhat different* group, while a greater percentage of international students fell into the *completely different* group.

A chi-square test to determine whether domestic and international teachers-in-training were distributed proportionally (that is, randomly) across the four language learning theory groups suggests that they were not ($\chi^2(3) = 12.74, p < 0.01$) (notice that I have included the appropriate degrees of freedom for this chi-square test in parentheses after χ^2 in this sentence, before providing the calculated chi-square value). The calculated chi-square value in this example (12.74) is higher than the critical chi-square value (11.34) for three degrees of freedom

Table 5.10 Observed distribution of reported change in second language learning theory for domestic and international students

	Domestic	International
Completely different	11 (22%)	15 (65%)
Somewhat different	19 (39%)	4 (17%)
Mostly the same	13 (27%)	2 (9%)
Exactly the same	6 (12%)	2 (9%)

Reproduced from Markham et al. (2017, Table 6)

($df = (R-1)(C-1) = 3*1 = 3$) when $\alpha = 0.01$ (you can look this up in Appendix C). Notably, as already described, domestic students are more likely to fall into the 'somewhat different' and 'mostly the same' categories compared to their international peers. This information is useful for individuals who work with domestic and international teachers-in-training—these two groups seem to respond differently to pedagogical interventions, such as exposure to certain theoretical approaches.

5.3 Looking Forward

You may have noticed that, until now, we have conducted hypothesis tests that involve at least one categorical variable. That is, to conduct a t-test, we needed at least one category to compare to a hypothesized value (e.g., study abroad returnees). For one-way ANOVA, we allowed more than two groups (e.g., study abroad applicants, virtual exchange applicants, and none-of-the-above), and in chi-square tests, both our variables were categorical (e.g., race/ethnicity and chosen intercultural learning experience). However, you may also want to explore the relationship between two continuous variables (e.g., international student enrollment numbers and the amount an institution spends on scholarships for international students). In this case, none of the hypothesis tests we have conducted thus far are appropriate. In the following chapter, we introduce *correlation*, which begins our discussion of exploring relationships between two continuous variables.

5.4 Practice Problems

1. Decide which statistical test (t-test, one-way ANOVA, or chi-square) is appropriate to address the following research questions:
   a. Are institutions located in rural areas (compared to non-rural areas) less likely to offer study abroad opportunities?
   b. Do institutions that focus primarily on STEM graduate training enroll more international students?
   c. What is the relationship between an institution's geographic location (rural, town, suburban, urban) and the number of international research collaborations (measured as number of publications co-authored with an international collaborator) that its faculty produce?
   d. What is the relationship between an institution's size (small, medium, large) and its likelihood of offering on-campus intercultural training opportunities?
   e. Do institutions that enroll international students have higher graduation rates compared to those that do not?
2. The data in the table below provide information about the number of international research collaborations of 18 faculty members located at research universities of three different sizes (small, medium, and large) (note that

5.4 Practice Problems

there are six researchers in each institution size category). Conduct a one-way ANOVA to test the null hypothesis that the mean number of international research collaborations of faculty members at institutions in each size category is the same.

Researcher	Small	Medium	Large	
1	4	5	8	
2	2	4	9	
3	1	1	4	
4	0	7	15	
5	0	8	12	
6	3	12	10	Grand mean
Mean	1.7	6.2	9.7	5.83

Note Data in this table are invented for illustrative purposes

  a. Calculate the sum of squares between and the mean of squares between.
  b. Calculate the sum of squares within and the mean of squares within
  c. Find the calculated F-statistic.
  d. Find the critical F-statistic ($\alpha = .05$).
  e. Do you reject or fail to reject the null hypothesis?
3. Using Sample Dataset #1 and the statistical software program of your choice, explore the null hypothesis that mean international student enrollment (intlstudents) at U.S. flagship institutions is the same for all institutional locales (rural, town, suburban, urban) ($\alpha = 0.05$).
4. The data in the table below provide information about the distribution of international student enrollment by region of origin (defined as the continent that a student is from) and the size of the institution of higher education where the student enrolled. Conduct a chi-square test to examine the null hypothesis that an international student's continent of origin is not related to the size of the institution where they enroll.

	Small	Medium	Large
Africa	356	746	987
Americas	1356	236	512
Asia	128	153	998
Australia/Oceania	345	321	413
Europe	98	352	732

Note Data in this table are invented for illustrative purposes

a. Find the row and column sums for this table.
b. Calculate the expected frequencies of international student enrollment by continent of origin and size of institution.
c. Find the calculated chi-square value.
d. Find the critical chi-square value when $\alpha = 0.05$.
e. Do you reject or fail to reject the null hypothesis?
5. Using Sample Dataset #2 and the statistical software program of your choice, explore the null hypothesis that whether a liberal arts college offers study abroad (studyabroad_offered) is not related to its geographic locale (rural, town, suburban, urban) ($\alpha = 0.05$).

Recommended Reading

A Deeper Dive

Urdan, T. C. (2017). One-way analysis of variance. In *Statistics in plain English* (pp. 113–132). Routledge.

Urdan, T. C. (2017). The chi-square test of independence. In *Statistics in plain English* (pp. 205–212). Routledge.

Additional Examples

Brunsting, N. C., Smith, A. C., & Zachry, C. E. (2018). An academic and cultural transition course for international students: Efficacy and socio-emotional outcomes. *Journal of International Students, 8*(4), 1497–1521.

Jin, L., & Schneider, J. (2019). Faculty views on international students: A survey study. *Journal of International Students, 9*(1), 84–99.

Lee, D., Allen, M., Cheng, L., Watson, S., & Watson, W. (2021). Exploring relationships between self-efficacy and self-regulated learning strategies of English language learners in a college setting. *Journal of International Students, 11*(3), 567–585.

Correlation 6

While the hypothesis tests involving at least one categorical variable explored in Chaps. 4 and 5 form an important foundation for quantitative analysis, many interesting questions for international educators involve relationships among continuous variables. For example, a researcher studying international student mobility might be concerned with the relationship between the number of students leaving their home countries to study elsewhere and variations in the GDP of students' home countries. In other words, on average, is the economic strength of a student's home country related to whether that student chooses to leave the country for education purposes? Another example might be the relationship between the number of internationally co-authored publications by an institution's faculty members and the amount of grant funding awarded to that institution. In other words, on average, do institutions whose faculty co-author with a greater number of international researchers also receive a greater amount of grant funding? Note that both these research questions ask about relationships between continuous variables, international student numbers and GDP in the first example and number of internationally co-authored publications and grant funding dollars in the second. The statistical test introduced in this chapter, correlation, helps us explore the relationship between two continuous variables and, importantly, rounds out the introduction of statistical concepts that we need before moving to more advanced analytic approaches, particularly regression.

6.1 Correlation: Main Ideas

The basic building block that we will use to begin our exploration of how continuous variables are related to each other is **correlation**, and specifically the **correlation coefficient**. The correlation coefficient provides a numerical summary of the strength of the relationship between two continuous variables. In this chapter, we will focus our attention on the **Pearson product-moment correlation**

coefficient, which is used to explore the relationship between two interval or ratio variables. This is the most common correlation coefficient used in social sciences research, although it is important to acknowledge that other kinds of correlation coefficients are available. Remember from Chap. 1 that interval and ratio variables are by nature numerical and that the distance between values of these variables is equal. Continuing with our faculty publications example, a faculty member who has co-authored 15 articles with international co-authors has exactly one more internationally co-authored article than a faculty member who has co-authored 14 articles with international co-authors.

Correlation coefficients can be positive or negative. A positive correlation coefficient indicates that two variables move in the same direction together—a higher value of variable A corresponds to a higher value of variable B and a lower value of variable A corresponds to a lower value of variable B (on average). Our international co-publications and grant funding example is an example of a correlation that might be positive—more international co-publications correspond to greater amounts of grant funding, while fewer international co-publications correspond to lower amounts of grant funding. This relationship is illustrated in Fig. 6.1, where larger values of grant funding roughly correspond to larger values in the number of internationally co-authored publications. In contrast, a negative correlation means that the two variables move in opposite directions. A higher value of variable A

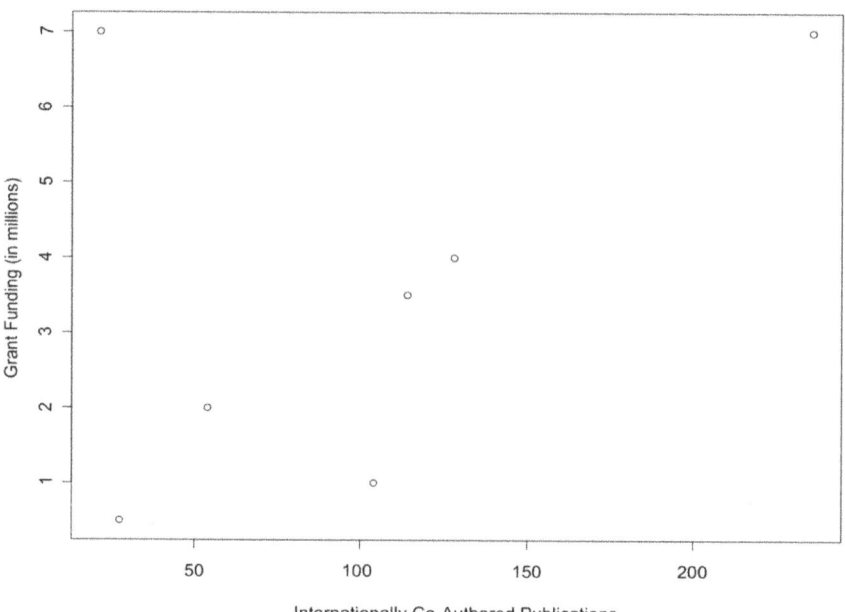

Fig. 6.1 Example of a positive correlation (internationally co-authored publications and grant funding). (*Note* Data in this figure are invented for illustrative purposes.)

6.1 Correlation: Main Ideas

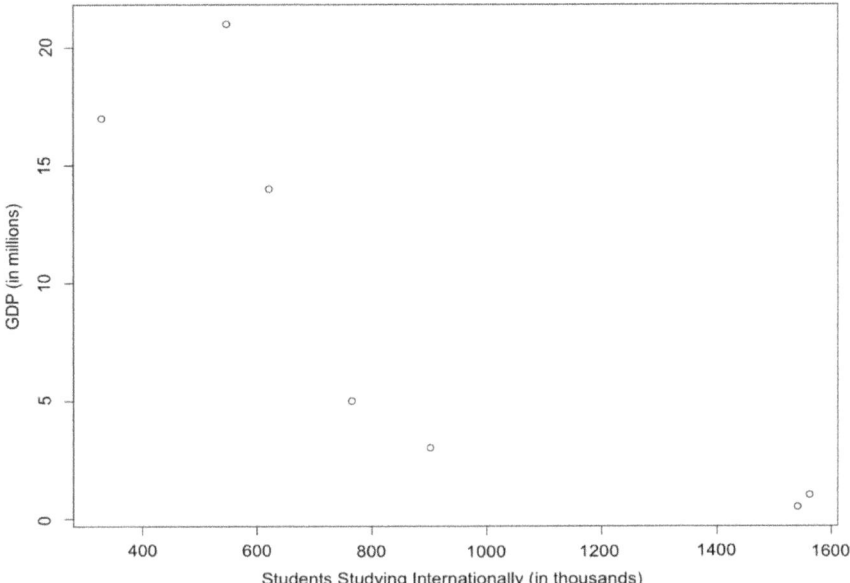

Fig. 6.2 Example of a negative correlation (students studying internationally and GDP). (*Note* Data in this figure are invented for illustrative purposes.)

corresponds to a lower value of variable B, and vice versa. Returning to our international student mobility and GDP example, we might think that students are, on average, more likely to remain in their home countries if their home countries are more economically prosperous. In this case, fewer students from countries with high GDPs study internationally. At the same time, more students from countries with low GDPs decide to leave their home countries to study abroad. This relationship is illustrated in Fig. 6.2. In this figure, higher values of GDP correspond to lower numbers of students studying internationally. Notice that the relationships in Figs. 6.1 and 6.2 are both linear, meaning that the average relationship between the two variables could be depicted (roughly) with a straight line. Correlation tells us very little about relationships that are not linear. A non-linear relationship would be present in our international student mobility example if the number of students studying abroad decreased as GDP increased only to a certain point and then, after that point, the number of students studying abroad increased as GDP grew (making a u-shape). While not discussed further in this chapter, we will return to non-linear relationships in the context of regression analysis in Chaps. 7 (Ordinary Least Squares Regression) and 8(Additional Regression Topics).

Numerically, correlation coefficients range in value from -1 to 1. This is because the correlation coefficient is a standardized version of the covariance, the amount of variance that two variables share, a concept that we will return to shortly. A value of 1 indicates a perfect positive correlation, meaning that the positive relationship between two variables holds for each and every unit of a study's sample, while a value of −1 indicates a perfect negative correlation, meaning that the negative relationship between two variables holds for each and every unit of a study's sample. These perfect correlations are rarely found in the real world. For example, the correlation between the two variables displayed in Fig. 6.1 is 0.42. Notice that while for most institutions in this sample, a higher number of internationally co-authored publications relates to higher amounts of grant funding, this is not always the case. Similarly, the correlation displayed in Fig. 6.2 is −0.84. While higher GDP generally suggests lower international student mobility, this is not the case for all the countries in this example.

Generally speaking, the closer a correlation coefficient is to −1 or 1, the stronger we say the correlation between two variables is. A correlation coefficient of 0, the midpoint between −1 and 1, indicates that there is no linear relationship between two variables. The closer a correlation coefficient is to 0, the weaker we say the correlation between two variables is. Such a non-relationship is illustrated in Fig. 6.3, where variables A and B exhibit no clear systematic relationship to one another at all. While the example in Fig. 6.3 is for illustrative purposes only, there is plenty of reason to expect no correlation between many variables that we might find in our datasets. For example, our dataset might contain information about the amount that a student paid for their study abroad experience and the number of courses they took in their first year of enrollment in college. There is no particular reason to believe that there is a correlation—positive or negative—between these two variables (although with some creative thinking, one could invent multiple

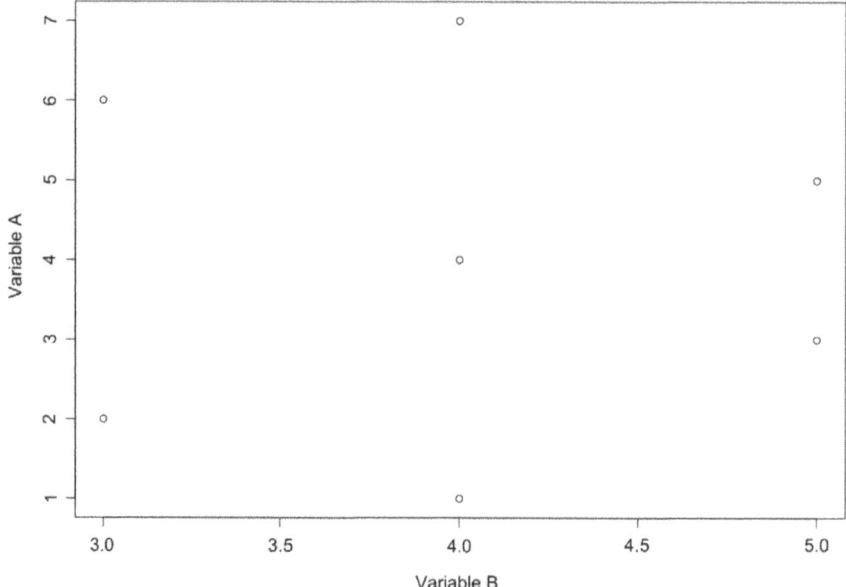

Fig. 6.3 Example of zero correlation. (*Note* Data in this figure are invented for illustrative purposes.)

explanations for a correlation in this situation—perhaps part-time students [who by definition take fewer classes at a time] come from lower socioeconomic status backgrounds and therefore participate in less expensive study abroad programs, or, perhaps these same students have full-time jobs and as a consequence can afford more expensive study abroad programs).

When first learning about correlation, it is reasonable to want to know specific numerical values that correspond to strong or weak correlations—for example, is a correlation coefficient of 0.63 close enough to 1 to be considered strong? Unfortunately, there is no straightforward answer to this question. You should use your theoretical and background knowledge on a given topic to thoughtfully interpret the strength or weakness of a correlation coefficient. For example, a correlation of 0.42 between GDP and international student mobility might be very strong if no relationship between the two variables was expected. At the same time, this correlation might be considered somewhat weak if previous research on this topic suggests that this relationship should be very strong. Correlations reported in prior research using the variables under study often can offer a good guide for interpreting correlation coefficients.

6.2 Calculating Pearson's Correlation Coefficient

Calculating the **covariance** between two variables is the first step in calculating Pearson's correlation coefficient. That is, just like the standard deviation is the standardized version of the variance of a variable, correlation is the standardized version of covariance, the joint variability of two variables. The formula for covariance (s_{xy}) is as follows:

$$s_{xy} = \frac{\sum(X - \overline{X})(Y - \overline{Y})}{n - 1}$$

Here, we first calculate the deviations from the mean for each individual observation of our two variables of interest, represented as X and Y. Next, we calculate the products of these deviations for each observation and then sum them. The final step in calculating covariance is dividing by $n - 1$, to account for degrees of freedom, just as we have done in other calculations. Notice how this formula is very similar to the formula for variance presented in Chap. 3 but instead of squaring the deviations of a single variable, covariance accounts for the product of deviations for two variables, X and Y.

Like variance, covariance suffers from a drawback that makes it impossible to compare relationships among variables that use different units of measurement. For this reason, we standardize covariances, turning them into correlation coefficients. This standardization is how we arrive at correlation coefficients that are always between -1 and 1. In notation, Pearson's correlation coefficient is represented using r for a sample and ρ for a population. The formula for calculating this coefficient divides the covariance between two variables by the product of

Table 6.1 Data corresponding to Fig. 6.2 (internationally mobile students and GDP)

Country	Variable X			Variable Y			$(X - \overline{X})(Y - \overline{Y})$
	GDP (in millions)			Students studying internationally (in thousands)			
	Raw data	$X - \overline{X}$	$(X - \overline{X})^2$	Raw data	$Y - \overline{Y}$	$(Y - \overline{Y})^2$	
A	21.00	12.21	149.19	547	−348.29	121,302.94	−4254.06
B	17.00	8.21	67.47	329	−566.29	320,679.51	−4651.63
C	14.00	5.21	27.19	621	−274.29	75,232.65	−1430.20
D	5.00	−3.79	14.33	765	−130.29	16,974.37	493.22
E	3.00	−5.79	33.47	902	6.71	45.08	−38.85
F	1.00	−7.79	60.62	1562	666.71	444,507.94	−5190.85
G	0.50	−8.29	68.65	1541	645.71	416,946.94	−5350.20
Mean	$\overline{X} =$ 8.79			$\overline{Y} =$ 895.29			
Sum			420.93			1,395,689.43	−20,422.57
SD			482.31			8.38	
/n − 1							−3403.76

Note Data in this table are invented for illustrative purposes

each variable's individual standard deviation (remember the formula for standard deviation in Chap. 3).

$$r_{xy} = \frac{s_{xy}}{(s_x)(s_y)}$$

Table 6.1 corresponds to the data displayed visually in Fig. 6.2 and provides the calculations needed to arrive at the correlation between international student mobility and GDP.

The numerator of our formula for calculating the correlation coefficient is the covariance between variables X and Y. This calculation requires three steps, summarized in the first equation above. First, the mean of variable X is subtracted from each individual value of variable X (this is what is happening in the column $X - \overline{X}$) and the mean of variable Y is subtracted from each individual value of variable Y ($Y - \overline{Y}$). If we focus on Country A in this example, we see that this country's GDP (variable X) is 21 (in millions). If we subtract the sample's average GDP (8.79), we get 12.21 ($X - \overline{X}$). In other words, Country A's GDP is around 12 million higher than the sample's average GDP. Likewise, for Country A, we see that the raw number of students studying internationally (variable Y) is 547 (given in thousands), and this number minus the average for all seven countries, 895.29, is −348.29 ($Y - \overline{Y}$). In other words, in Country A, approximately 350 fewer students went abroad to study compared to this sample's average. This first step in calculating covariance accounts for deviations from the mean of our two variables of interest. In the second step, these deviations from the mean of the two variables are multiplied, so we end up with the product of the deviations of both variables $(X - \overline{X})(Y - \overline{Y})$. For Country A, this value is −4254.06. The final step in our covariance calculation is to sum these products for the whole dataset and divide by $n - 1$. In this case, the sum of products is $-20{,}422.57$ and $n - 1 = 6$. Our final calculated value for the numerator, -3403.76, is bolded at the bottom of the last column of Table 6.1 and is the covariance between GDP (variable X) and number of students studying internationally (variable Y). This numerator describes the relationship between variables X and Y, but its value can vary dramatically depending on the scales used to measure our variables of interest.

To standardize the covariance between two variables, we divide the covariance we just calculated by the product of the standard deviation of the two variables, provided in the next to last row in Table 6.1. The product of these two values is approximately 4041.76. If we divide the covariance between Variables X and Y (-3403.76) by this value, we arrive at a correlation coefficient of -0.84.

6.3 Additional Correlation Calculations

In addition to the correlation coefficient itself, two other calculations are also often useful for researchers comparing two continuous variables, the coefficient of determination and significance testing of the correlation coefficient.

6.3.1 The Coefficient of Determination

The **coefficient of determination** provides researchers with information about the percentage of variance that is shared between two variables that are correlated with one another. In our international student mobility and GDP example, we might be interested in understanding not only how these two variables are related to one another, but also how much about one variable do we know if we have information about the other variable. In other words, can we explain something about international student mobility if we know something about GDP (and vice versa)? In statistics, we often discuss the amount of explained variance between two variables or how much shared variance there is between two variables (more on explained or shared variance when we get to Ordinary Least Squares regression in Chap. 7). The coefficient of determination provides this information. Luckily, the coefficient of determination is also quite easy to calculate—it is simply the square of the correlation coefficient. In our running example, the correlation coefficient between number of students studying internationally and GDP was −0.84. If we square this number, we get 0.71—indicating that 71% of the variance between these two variables is shared.

6.3.2 Significance Testing

Finally, in addition to knowing the numerical correlation between two variables and how much of the variance two variables share, we might also want to know if a correlation is significant at a standard level (perhaps setting alpha to 0.05). To do this, we conduct a type of *t*-test that examines whether the correlation coefficient *r* is significantly different from a hypothesized population correlation coefficient of zero. The null (H_0) and alternative (H_A) hypotheses for this test are as follows (remember that ρ corresponds to the population correlation coefficient):

$$H_0 : \rho = 0$$

$$H_A : \rho \neq 0$$

I wrote these two hypotheses as mathematical equations rather than words to illustrate a different way of defining the null and alternative hypotheses than what we have seen in previous chapters. We could alternatively write these hypotheses out in words:

H_0 : The population correlation coefficient is equal to 0.

H_A : The population correlation coefficient is not equal to 0.

Notice that this is a two-tailed hypothesis, so we reject the null hypothesis if the correlation coefficient is significantly higher or lower than zero. We use the following formula to find our calculated *t*-value:

$$t = \frac{r - \rho}{s_r}$$

where r is the sample correlation coefficient, ρ is the population correlation coefficient (hypothesized to be zero), and s_r is the estimated standard error of the sample correlation coefficient. Following our null hypothesis, we set ρ equal to zero and are left with the correlation coefficient divided by the estimated standard error (s_r). s_r is calculated as follows:

$$s_r = \sqrt{\frac{(1 - r^2)}{(n - 2)}}$$

where n is the number of observations in our sample. We subtract 2 from *n* in the denominator in this formula, which represents one degree of freedom for each of the two variables in our analysis.

In our running example with a correlation coefficient of −0.84, we can calculate the standard error as follows (note that −0.84 squared is 0.71, and our sample size is N = 7 countries):

$$s_r = \sqrt{\frac{(1-0.71)}{(7-2)}} = 0.24$$

We can then calculate our *t*-value:

$$t = \frac{-0.84}{0.24} = -3.50$$

The *t*-table in Appendix A indicates that with five degrees of freedom (7 − 2 = 5), the critical *t*-value for a two-tailed hypothesis test is 2.5706 ($\alpha = 0.05$). Remember that for a two-tailed hypothesis test, we compare the absolute value of our calculated test statistic to the critical value (the absolute value of −3.50 is 3.50). In this case, 3.50 is greater than the critical value, and so we reject the null hypothesis, finding a significant correlation coefficient ($p < 0.05$).

While researchers are often excited to find a significant correlation, note that this test of significance is not especially useful from a practical standpoint. Significant correlations are not necessarily important once other variables are considered (e.g., lower GPD might actually be indicative of less governmental funding for higher education, which may be the more important variable to account for in the case of our running example), and, of course, correlation is not the same as causation (more on this point later). In fact, it is quite common for researchers to find spurious correlations between variables in their datasets. Spurious correlations occur when two variables at first appear to be related to one another but, in reality, are not at all connected. International education researchers should interpret significant correlations with caution—these correlations must be interpreted in the context of prior research, theory, and maybe most importantly, common sense.

6.4 Examples from Sample Data and the Literature

This section summarizes two examples of correlation that help to illustrate how this statistical concept works in real life, one using one of this book's sample datasets and another from a published research study. Although our running examples of correlations between international student mobility and GDP on the one hand and international co-authorship and grant funding on the other provide us with important contexts for understanding correlation, invented datasets such as these that include only seven observations are not particularly realistic.

The first example expands on the correlation calculations presented in this chapter using this book's Sample Dataset #1, which contains data corresponding to 50 US flagship institutions (one for each state). Here, we explore two of this dataset's continuous variables: international student enrollment and an institution's total expenditure on instruction. Figure 6.4 displays the relationship between these two variables visually. Given what we see in this figure, we might expect the correlation between these two variables to be positive.

Indeed, this correlation coefficient is $r = 0.81$. If we square this value of r, we find that the coefficient of determination for these two variables is 0.66, meaning

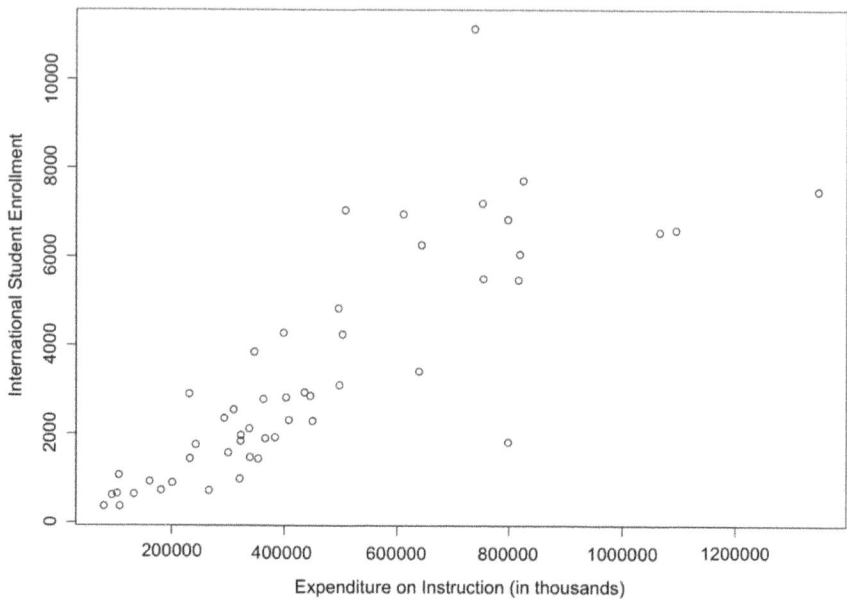

Fig. 6.4 Correlation between international student enrollment and total expenditure on instruction at U.S. Flagship Institutions from the 2015–16 academic year. *(Data source:* Sample Dataset #1 - US National Center for Education Statistics, Integrated Postsecondary Education Data System)

that they share around 66% of their variance. A hypothesis test exploring whether this correlation coefficient is significantly different from zero yields a *t*-value can be calculated as follows :

1. Calculate the standard error of the sample correlation coefficient:

$$s_r = \sqrt{\frac{(1-r^2)}{(n-2)}} = \sqrt{\frac{(1-0.66)}{(50-2)}} = \sqrt{\frac{0.34}{48}} = \sqrt{0.007} = 0.08$$

2. Use this standard error in the formula for *t*:

$$t = \frac{r - \rho}{s_r} = \frac{0.81 - 0}{0.08} = 10.13$$

3. Compare this calculated *t*-value with the critical *t*-values in Appendix A.

Because this calculated *t*-value is considerably larger than the critical *t*-values in Appendix A corresponding to both 40 and 60 degrees of freedom (remember that we have $n - 2$, or 48 in this case, degrees of freedom), we reject the null hypothesis that the correlation coefficient is significantly different from zero. This finding suggests that institutions that spend more money on instruction enroll more

6.4 Examples from Sample Data and the Literature

international students, on average. However, note that this relationship cannot be interpreted as causal—raising instructional expenditure does not necessarily cause additional international student enrollment. In fact, the reverse relationship could also be true—institutions might spend more on instruction *because* they enroll more international students.

As a second example of correlation, Hendrickson and colleagues (2011) explore the relationships between measures of the quality of international students' friendship networks and students' feelings of social connectedness. Participants in this study were 86 international students studying at the University of Hawaii who completed a survey that, among other information, collected students' responses to items that measured social connectedness, homesickness and contentment, and satisfaction with life (all measured on a five-point Likert scale). Students also completed a friendship network grid, which asked them to indicate where their friends were from and what the strength of those relationships were (rated on a scale from 1 to 10). Students' friendships were disaggregated into three categories: co-national (from the same country as the student), multi-national (from other foreign countries), and host national (from the host country). As part of their analysis, the authors use correlation to explore the relationships between students' feelings of social connectedness, homesickness, contentment, satisfaction with life, and the strength of their relationships with other students of varying origins (co-national, multi-national, or host national) (among other measures not discussed here). The Pearson's correlation coefficients corresponding to their results are reproduced in Table 6.2 (adapted from Table 2 in Hendrickson et al. [2011]). This table is an example of what is often referred to as a correlation matrix. A correlation matrix displays all the correlations among all the possible pairs of variables in a given set.

Notice in this table that any measure correlated with itself is equal to 1 (e.g., in the first row of the table, satisfaction is correlated with satisfaction at $r = 1$), a logical, but trivial, result. More substantively, several significant correlations ($p < 0.05$ or lower) were found in Hendrickson et al.'s study. For example, students' contentment and satisfaction with life were correlated at $r = 0.42$, while social connectedness and satisfaction with life were correlated at $r = 0.34$. The strengths of students' friendship relations were also significantly correlated with one another. For example, the strength of multi-national friendships was correlated with the strength of host national friendships ($r = 0.30$). These correlations provide us with some initial insight into the relationships between international students' well-being and the friendships that they form. For example, international students who are more content are, on average, also more satisfied with life. Students who form strong relationships with host nationals also tend to share strong relationships with individuals from other countries. However, these results cannot tell us anything about the direction of the relationship. For example, these results cannot tell us if contentment causes satisfaction with life or vice versa. This is a point we will return to again over the course of the next few chapters.

Table 6.2 Correlations between satisfaction, contentment, homesickness, connectedness and host national friendship strength, co-national friendship strength, and multi-national friendship strength.

	Satisfaction	Contentment	Homesickness	Connectedness	Host national strength	Co-national strength	Multi-national strength
Satisfaction	1						
Contentment	0.42**	1					
Homesickness	−0.13	−0.21	1				
Connectedness	0.34**	0.63**	−0.08	1			
Host national strength	−0.03	0.00	−0.08	0.02	1		
Co-national strength	−0.19	−0.10	0.16	−0.06	0.07	1	
Multi-national strength	0.01	0.06	0.01	0.18	0.30**	0.28*	1

* $p < 0.05$, ** $p < 0.01$
Adapted from Table 2 in Hendrickson et al. [2011], p. 289

6.5 Correlation and Causation

To round out this chapter, we briefly turn our attention to the issue of correlation and causation. While our discussion until now has been descriptive, describing the *relationship* between two variables, many questions that international education researchers ask at least implicitly refer to causal relationships. That is, does a lower GDP *cause* students to leave their home countries to study abroad? Or, does an increase in internationally co-authored articles *cause* a faculty member to be awarded larger amounts of grant funding? Correlation alone cannot provide evidence that one particular variable causes another. If we observe a relationship between internationally co-authored publications and grant funding, this might be evidence of a true causal relationship, but many other factors might also explain why faculty members who tend to co-author internationally also tend to receive more grant funding. For example, these faculty members may be more likely to co-author in general, not just internationally, and it is the collaborative aspect of these publications that leads to more grant funding. Another explanation might be that faculty members who co-author internationally produce a greater amount of research, both international and domestic, and this is the reason why they receive greater grant funding. While correlation does not imply causation, correlation between two variables is a necessary condition for more complex analyses that are able to speak to potential causal relationships. If two variables do not correlate at all, then a causal relationship from one to the other is likely not present. More advanced statistical methods that can be used to investigate causal relationships are introduced in Chap. 9, but we will explore important components of these research designs in the chapters on regression, Chaps. 7 and 8.

6.6 Practice Problems

1. The following table provides the raw data used to produce Fig. 6.1, which visually displayed the relationship between internationally collaborative publications and grant funding.

Institution	Internationally collaborative publications	Grant funding amount (in millions)
A	27	0.5
B	236	7
C	104	1
D	54	2
E	128	4
F	114	3.5
G	21	7
Mean	97.71	3.57

Note Data in this table are invented for illustrative purposes

Use this table to calculate the following three pieces of information about this correlation by hand:

a. The correlation coefficient
b. The coefficient of determination
c. Whether the correlation coefficient is significantly different from zero

2. Sample Dataset #1 contains information about international student enrollment numbers (intlstudents) and total enrollment (totalenroll). Figure 6.5 visualizes these two variables.

Relying on Figure 6.5, do you expect the correlation between these two variables to be positive or negative? Why?

3. Using this dataset and your statistical software program of choice, calculate the following pieces of information about the relationship between international student enrollment numbers and a flagship institution's total enrollment:
a. The correlation coefficient
b. The coefficient of determination
c. Whether the correlation coefficient is significantly different from zero

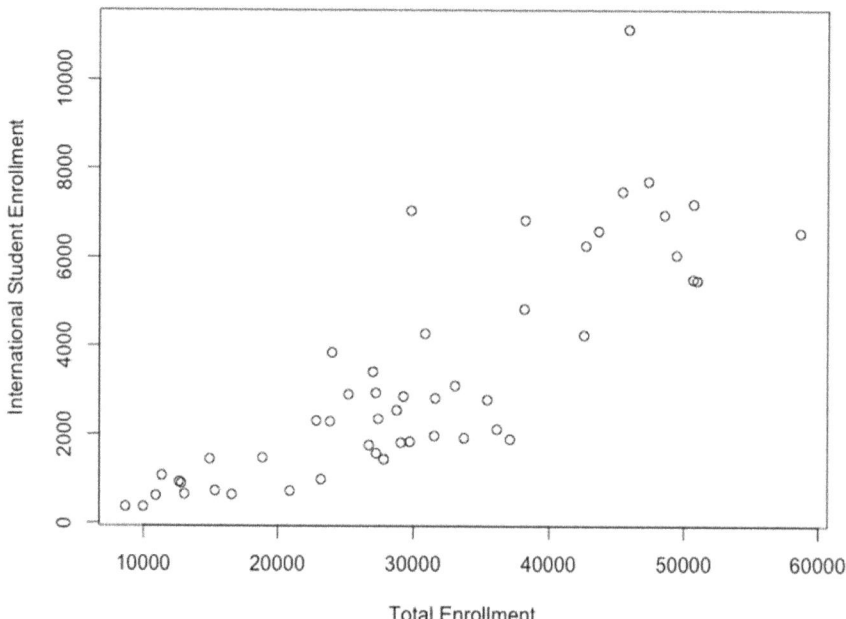

Fig. 6.5 Correlation between international student enrollment and total enrollment at U.S. Flagship Institutions from the 2015–16 academic year. (*Data source*: Sample Dataset #1 - US National Center for Education Statistics, Integrated Postsecondary Education Data System)

Recommended Reading

A Deeper Dive

Wheelan, C. (2013). Correlation: How does Netflix know what movies I like? In *Naked statistics: Stripping the dread from data* (pp. 58–67). Norton.

Urdan, T.C. (2017). Correlation. In *Statistics in plain English* (pp. 165–182). Routledge.

Additional Example

Güzel, H., & Glazer, S. (2019). Demographic correlates of acculturation and sociocultural adaptation: Comparing international and domestic students. *Journal of International Students, 9*(4), 1074–1094.

Ordinary Least Squares Regression

7

Chapter 4 (Hypothesis Testing), Chap. 5 (One-way ANOVA and the Chi-square Test of Independence), and Chap. 6 (Correlation) helped us to explore the relationships between two variables in a dataset and make inferences about a population based on sample data. The way we explored relationships depended on the types of variables we were comparing. For example, when we were comparing two groups (represented with a dichotomous variable, such as students who applied to study abroad and those who did not) in terms of a continuous variable (such as a student's GPA), we used a *t*-test. We used correlation to explore the relationship between two continuous variables, such as international student enrollment and d an institution's expenditure on instruction. While you were reading about these statistical tests, you may have noticed a shortcoming of all of them—in each case, the variable that we were interested in, such as international student enrollment, probably depended on more than one other variable—e.g., not only expenditure on instruction, but also other institutional characteristics, such as research institution status or geographic location. This chapter builds on the concepts behind the two-variable hypothesis tests we have explored thus far to introduce one of the most commonly used analytic approaches in education research—regression, which addresses this shortcoming.

Regression is a popular statistical technique because it is so versatile. It helps us make predictions about outcomes, explore the strength of the relationship between two variables, and isolate the relationship between two variables even when more than one variable is involved in predicting a certain outcome (sometimes called *partialling out* relationships or *controlling for* additional variables, a concept we will return to later in this chapter). Many researchers use regression for what is called *causal inference*, meaning that they want to draw conclusions about whether a change in one variable *causes* a change in another. Accounting for other variables that might also cause that same change is important in this situation, since doing so can rule out other competing hypotheses about what might be causing a specific change (for more on causal inference, see Chap. 9). However, regression can also be used for descriptive analysis, like we have been doing. In this sense, it can be used to describe the relationships among variables in a particular dataset without any reference to causality. We might, for example, want to describe the number of students who will study abroad from a particular institution based on a series of institutional characteristics, such as the amount of financial aid that students receive, the demographic characteristics of the students who attend, and the size of the institution (perhaps measured in terms of enrollment numbers). An important characteristic of regression is that it is flexible in the kinds of variables that can be included in a single analysis. For example, we might explain international student enrollment using both expenditure on instruction (a continuous variable) *and* whether an institution is a research university (a dichotomous, categorical variable).

In this chapter, we will specifically explore two regression approaches: **simple linear regression** and **multiple linear regression**. The next chapter (Chap. 8: Additional Regression Topics) provides additional analytic details about regression. A foundation in linear regression is essential before we move to those topics. In brief, both simple linear regression and multiple linear regression (often referred to by their typical estimator, ordinary least squares regression) focus on explaining a variable that is continuous (e.g., international student enrollment) —the variable of interest. Simple linear regression uses one variable to predict the variable of interest while multiple linear regression allows researchers to account for several explanatory variables in a single statistical model.

All regression models contain two types of variables. The first is the variable that you are trying to explain—international student enrollment in our running example. This variable is often referred to as the **outcome variable** or the **dependent variable**. In standard statistical notation, a dependent variable is represented as y and it is displayed along the y-axis (the vertical axis) when graphing data. The second kind of variable is the variable used to do the explaining—such as expenditure on instruction. These variables are often referred to as **explanatory variables**, **predictor variables**, **covariates**, or **independent variables**. In standard statistical notation, independent variables are represented with x and are represented along the x-axis (the horizontal axis) in graphs. The logic behind some of these terms will become clear as we work through the underlying concepts of regression modeling. What is important to understand for now is that, unlike the analyses we have seen thus far (such as correlation or *t*-tests), regression assumes a direction for the relationship between two variables. One variable predicts another (regardless of whether the relationship is causal or not).

7.1 Simple Linear Regression

The goal of simple linear regression is to find the line that best depicts the relationship between two variables. Here, it helps to have a visual to clarify. We return to the correlation between international student enrollment and total expenditure on instruction depicted back in Fig. 6.4, reproduced here as Fig. 7.1. This figure is an example of what is called a **scatterplot**, a way of visualizing data wherein

7.1 Simple Linear Regression

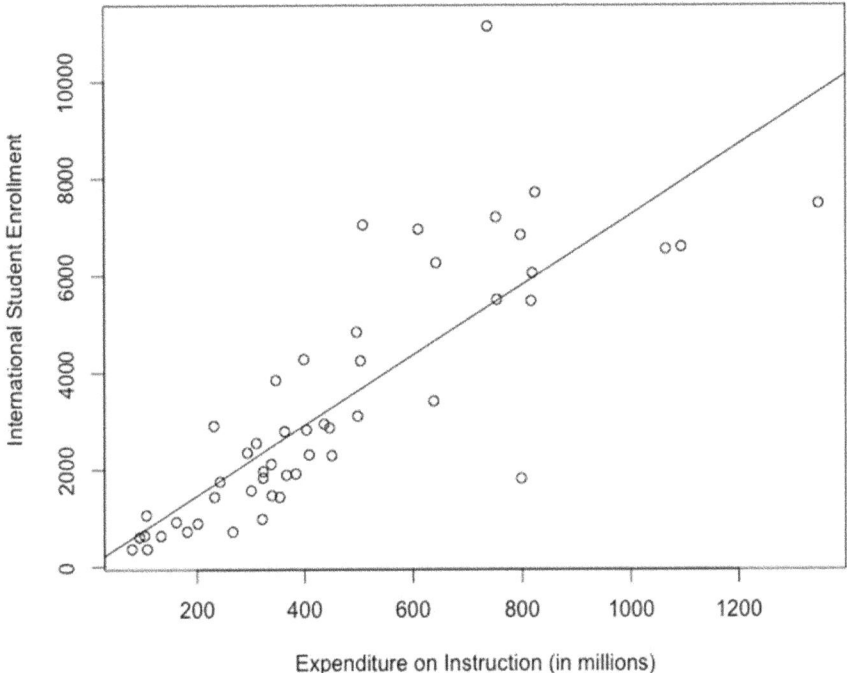

Fig. 7.1 Linear relationship between international student enrollment and total expenditure on Instruction at U.S. Flagship Institutions from the 2015–16 academic year. (*Data source:* Sample dataset #1 - US National Center for Education Statistics, Integrated Postsecondary Education Data System)

each data point is represented along the x- and y-axes of a graph, each of which represents a distinct variable. This kind of plot is useful when we want to explore whether a relationship might exist between two variables. Figure 7.1 contains one key addition that makes it different from Fig. 6.4 -- Fig. 7.1 contains what is called a **regression line**. This line is roughly drawn through the middle of the data points and depicts the best estimate of the average relationship between the two variables. For the purposes of this example, we assume that expenditure on instruction predicts international student enrollment, and not the other way around (although it is quite possible that the relationship goes the other way—revenue from tuition that international students pay could be invested in instruction).[1]

The **slope** of the regression line is the ratio of the vertical change to the horizontal change and gives us the predicted average increase in international student

---

[1] Remember that the dataset that corresponds to this analysis includes flagship institutions from the 50 states in the 2015–16 academic year. These data were extracted from the National Center for Education Statistics' Integrated Postsecondary Education Data System (IPEDS) (see nces.ed.gov/ipeds for additional information).

enrollment (the dependent variable) when expenditure on instruction (the independent variable) increases by one unit. In other words, the regression line indicates the predicted value of a given dependent variable (international students) at a certain value of an independent variable (expenditure on instruction). Note that in this example, expenditure on instruction is measured in millions of U.S. dollars, so a one unit increase in expenditure on instruction is US$1,000,000. Researchers will often make a transformation like this one, which involves simply dividing the variable by a certain value (1,000,000 in this case), to make interpretation easier. When I conducted a simple linear regression with international student enrollment as my dependent variable and expenditure on instruction as my independent variable, I found that a US$1,000,000 increase in expenditure on instruction was associated with an average increase of seven enrolled international students. Note that this relationship is not necessarily causal—just because an institution increases expenditure on instruction does not mean that more international students will enroll—but this pattern is the tendency present in the data. In the rest of this section, I explain how I arrived at this answer.

7.1.1 Functional Form

If the goal of regression is to find the line that best depicts the relationship between two variables, we find ourselves in a place where we need to be a bit more specific about the properties of this line. That is, there are multiple lines that can be drawn through our data points, so how do we choose just one? In Fig. 7.1, I depicted this line as a straight line that approximately cuts through the middle of the data. The slope of this line represents the relationship between expenditure on instruction and international student enrollment. Notice that only a couple of data points fall directly on the line, while the other points are scattered at varying distances from the line's path. That is, this line depicts the *average* relationship between the two variables. I could potentially improve how well this line represents my data—thus improving the fit of my regression—if I added twists or curves to it. For example, I could make this line a bit more u-shaped, as in Fig. 7.2 so that more data points are closer to the line.

Another option would be to draw a line that, while substantially more complicated, connects *all* the data points and depicts an exact relationship, as in Fig. 7.3.

What these three examples illustrate are really questions of **functional form**. When we talk about a regression model's functional form, we are referring to the line that we have decided best depicts the relationship between two variables. While all three of these options (Figs. 7.1, 7.2, and 7.3) are possible functional forms that describe the relationship between international student enrollment and expenditure on instruction, they come with tradeoffs in terms of complexity. The straight linear relationship is the simplest and most easily interpreted (Fig. 7.1) while the exact relationship (Fig. 7.3) is the most complex. Indeed, the relationship depicted in Fig. 7.3 is so complex that it is not especially useful—it no longer

7.1 Simple Linear Regression

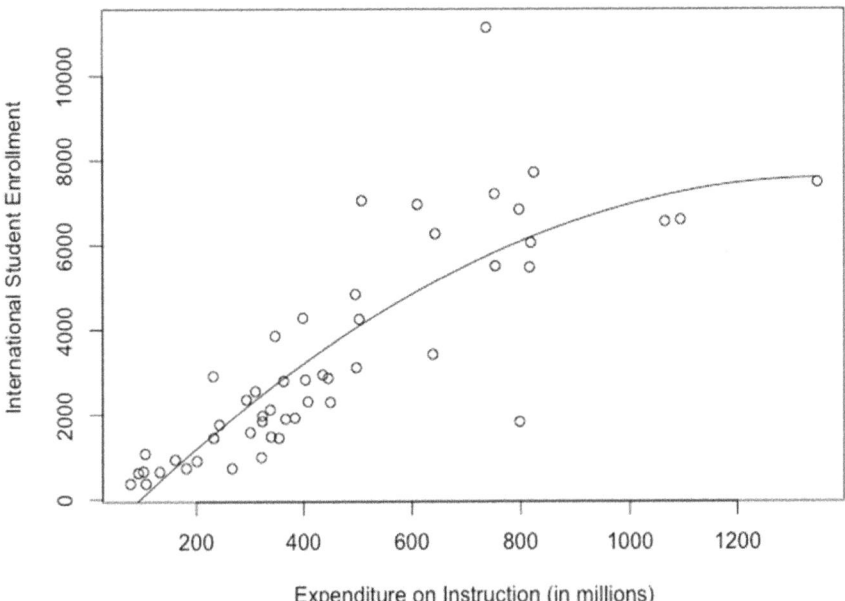

Fig. 7.2 U-shaped linear relationship between international student enrollment and total expenditure on instruction at U.S. flagship institutions from the 2015–16 academic year. (*Data source:* Sample dataset #1 - US National Center for Education Statistics, Integrated Postsecondary Education Data System)

depicts the *average* relationship between the two variables. In statistics, we would refer to the regression line depicted in Fig. 7.3 as an example of *overfitting*, meaning that it is so specific to the unique data points that it is no longer useful in representing the average data point or making predictions about units that are not in the dataset. While the curved line depicted in Fig. 7.2 may hold some value in exploring the relationship between international student enrollment and expenditure on instruction (and we will return to this functional form in the next chapter), the slope of this line is not constant (i.e., it is steeper at some places than others), and so this regression line is a bit complicated. For the sake of illustration, we will stick with the simplest (straight) line depicted in Fig. 7.1 to define the regression line in this chapter. In practice, this is what many researchers do on a regular basis because a straight line is a general functional form that is appropriate in many contexts.

7.1.2 Ordinary Least Squares

To estimate a linear regression, we lean on a statistical theorem called the Gauss-Markov Theorem. While the details of the Gauss-Markov Theorem are beyond the scope of this book (interested readers should certainly explore these details in a

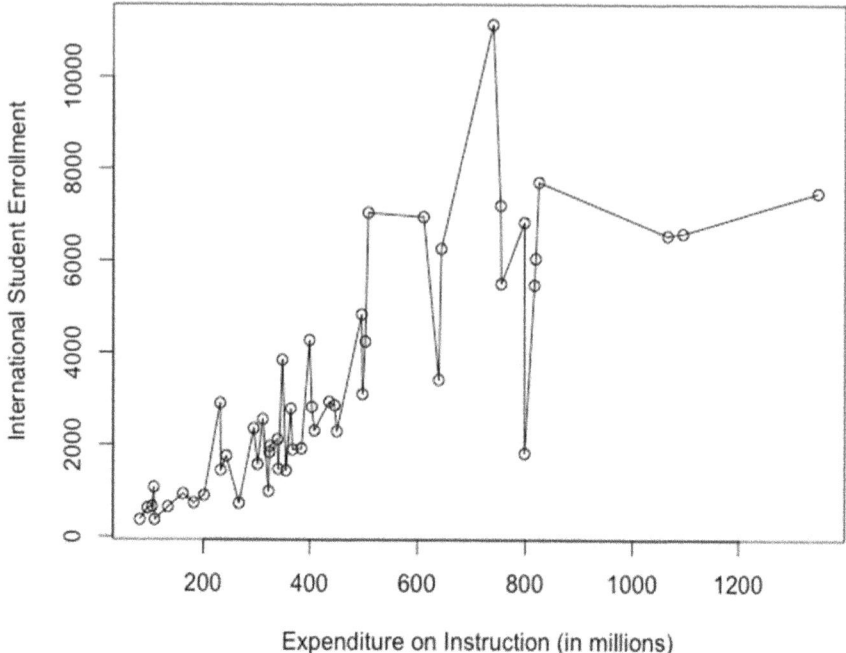

Fig. 7.3 Exact linear relationship between international student enrollment and total expenditure on Instruction at U.S. Flagship Institutions from the 2015–16 academic year. (*Data source:* Sample dataset #1 - US National Center for Education Statistics, Integrated Postsecondary Education Data System)

more advanced statistics book), this theorem shows that, under specific assumptions, an unbiased and efficient regression line is found using a method called **ordinary least squares** (or OLS).[2] Conceptually, ordinary least squares (OLS)

---

[2] These assumptions include: (1) correct functional form (we assume that the relationship between the two variables is linear in some way—but note that this line does not have to be straight. In more technical terms, we assume that the relationship depicted in the regression line is linear in its parameters), (2) random sampling, meaning especially that units in the dataset are not related to one another in some way (e.g., students who study abroad in the same program might be related to each other in ways not captured in your dataset), that is, there is no autocorrelation among units (see Chap. 1 for more information about sampling strategies and why randomization is important), (3) non-collinearity (when we use more than one independent variable to predict a dependent variable, these two independent variables are not perfectly correlated with one another. That is, they measure different things—see the section in this chapter on Multiple Linear Regression for additional details), (4) homoscedasticity (the error term is the same across all values of the independent variable), and (5), exogeneity, which is only necessary when regression is being used for causal inference (independent variables are not dependent on the dependent variable—that is, in this example, instructional expenditures are not dependent on international student enrollment. In other words, the relationship between the two variables only goes one way).

7.1 Simple Linear Regression

regression finds the regression line that minimizes the sum of the squared errors between the observed values and the regression line—the predicted values. We visited the concept behind summed squared errors—we called them squared deviations—when we calculated variance in Chap. 3. The concept here is exactly the same only this time instead of measuring deviations from the mean, we want to know the difference between the observed values of an outcome variable and the predicted values for this variable that are found along the regression line (that is, we want to know the *error* in the regression line). This concept is depicted visually in Fig. 7.4, which again shows the relationship between international student enrollment and instructional expenditure. The dotted line that connects data point A (representing one of the institutions in our dataset) to the regression line is the amount of error between the actual number of international students enrolled at this institution (6955) and the number of international students predicted to enroll at this institution based on its instructional expenditure (around 4400). In this case, a regression model based on only instructional expenditure underpredicts the number of students enrolled at institution A. The overarching goal of OLS regression is to draw the regression line that minimizes the squared error for all data points.

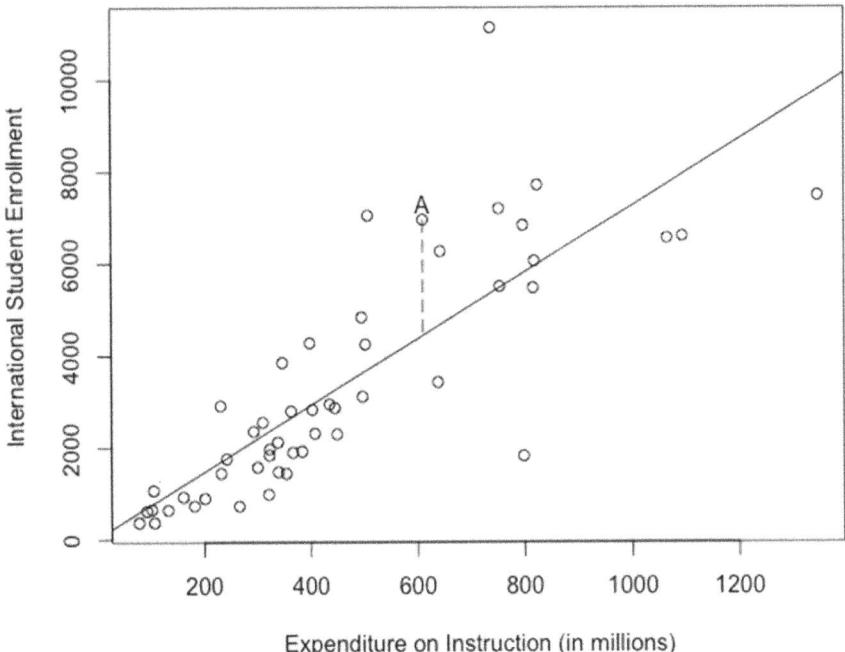

Fig. 7.4 Illustration of error between actual data point and regression line (represented as the dotted line connecting point a to the regression line). (*Data source:* Sample Dataset #1 - US National Center for Education Statistics, Integrated Postsecondary Education Data System)

7.1.3 Calculating a Simple Linear Regression

To calculate a simple linear regression using ordinary least squares, we use the following formula:

$$\hat{Y}_i = a + bX_i$$

Here, $\hat{Y}_i$ represents the predicted value of our dependent variable (international student enrollment), X_i is a particular value of an independent variable (the subscript i indicates that X has a value for each *i*ndividual observation in the dataset), and a and b represent the **intercept** and slope of our regression line, respectively. The intercept is the place where the regression line crosses the vertical (y-)axis (in Fig. 7.4, this is the number of international students that would be predicted to enroll at a particular institution if zero dollars were spent on instructional expenditure). The slope is, as already mentioned, the ratio of the vertical change to the horizontal change in the regression line (you might remember the phrase *rise over run* from a previous math class). Another term used to refer to the slope is the **regression coefficient**, and this is the term you will hear researchers use most frequently. The equation for finding the slope of a regression line is as follows:

$$b = \frac{\left(\frac{1}{n-1}\right) \sum (X_i - \overline{X})(Y_i - \overline{Y})}{\left(\frac{1}{n-1}\right) \sum (X_i - \overline{X})^2}$$

Notice that, while this equation looks complex, we have already seen the formulas in the numerator and denominator of this equation. The numerator is the covariance of the independent variable (X) and the dependent variable (Y) and the denominator is the variance of the independent variable (X). Substantively, this means that our regression coefficient is computed with the level of association between X and Y (the covariance) scaled by the variation in X (the variance). Notice that in this equation, the terms $\frac{1}{n-1}$ in the numerator and the denominator, which account for degrees of freedom, cancel each other out.

Once we calculate the slope of a regression equation, calculating the intercept is relatively straightforward. The intercept is simply the average of the dependent variable (international student enrollment) once the independent variable has been taken into account (that is, *partialled out* in more technical terms). The equation for calculating the intercept is as follows:

$$a = \overline{Y} - b\overline{X}$$

In this equation, $\overline{Y}$ is the mean of the dependent variable, b is the slope that we just calculated, and $\overline{X}$ is the mean of the independent variable.

Regression analysis can be calculated by hand. However, the simple calculations presented here can become unwieldy when the dataset being analyzed is large, a characteristic of many datasets. For this reason, researchers usually rely on statistical software programs for regression. When I conducted a simple linear

7.1 Simple Linear Regression

regression analysis for the relationship depicted in Fig. 7.4, where international student enrollment was the dependent variable and expenditure on instruction was the independent variable, my results indicated that the intercept of the regression line was approximately 17 and the slope of the regression line was approximately 7. Pairing words with the coefficient (or slope) ($b = 7$) on expenditure on instruction (X), I would say that a US$1 million increase in expenditure on instruction was associated with an average increase of 7 enrolled international students. In other words, if an institution increased expenditure on instruction by a million dollars, we would predict seven additional international students to enroll, on average. Again, this does not mean that the increase in instructional expenditure caused the increase in international student enrollment. It only means that the two variables are positively related descriptively.

The regression model that we just estimated is represented in equation form as follows:

$$\hat{Y}_i = 17 + 7X_i$$

Besides describing a general pattern in the data, regarding international student enrollment and instructional expenditure, this equation can be used to predict the number of international students that an institution will enroll ($\hat{Y}_i$) based on different levels of expenditure on instruction (X_i). That is, we can plug in different values of X_i to obtain **predicted values** for $\hat{Y}_i$. Table 7.1 illustrates a few of these calculations.

Table 7.1 suggests that, if an institution spent no money on instruction, we would still expect 17 international students to enroll. In this case, the intercept of the regression equation makes some sense to the extent that we think an institution could get away with spending nothing on instruction. However, sometimes intercept values are not particularly logical, especially when zero is not a legitimate option for the independent variable. (Note, however, that even in these cases the intercept is important for positioning the regression line correctly in a plot) If an institution spent only US$5 million on instruction, we would expect the number of international students enrolled to be 52. This prediction increases to 87 international students when expenditure on instruction is increased to US$10 million

Table 7.1 Predicted values of international student enrollment based on expenditure on instruction

Expenditure on Instruction (X) (in millions)	Calculation	Predicted International Student Enrollment ($\hat{Y}$)
US$0	$\hat{Y} = 17 + 7(0)$	17
US$5	$\hat{Y} = 17 + 7(5)$	52
US$10	$\hat{Y} = 17 + 7(10)$	87
US$20	$\hat{Y} = 17 + 7(20)$	157
US$1200	$\hat{Y} = 17 + 7(1200)$	8417

and to 157 when expenditure on instruction is US$20 million. When expenditure on instruction is US$1200 million (around the maximum in this dataset), we predict the institution to enroll around 8417 students. Again, these values refer to expected or predicted values. They do not imply that a specific increase in instructional expenditure necessarily causes a specific increase in international student enrollment.

7.1.4 Regression Residuals and Error

It is important to remember that in the regression equation we just presented ($\hat{Y}_i = a + bX_i$), we were dealing with *predicted* rather than *actual* values of our dependent variable (Y_i). The differences between actual values (also called **observed values**) and predicted values are called the **errors** or **residuals** (the variance in the dependent variable that gets "left over") in a regression model. Recall data point A in Fig. 7.4. In the case of this institution, the regression model underpredicted the number of international students enrolled. The distance between the actual data point and the regression line represents the error in the regression model for this particular institution. For other institutions, the number of international students was overpredicted, also resulting in error in our predictive model.

In reality, all estimated regression equations contain some error regarding the observed values of the dependent variable. This is a necessary side effect of choosing a simpler, but more practically useful, functional form (notice that the line depicted in Fig. 7.3 has no error but is also fairly useless for either explanation or prediction because the line is so complex). This error is often represented as e when a regression equation refers to observed values of Y rather than predicted values, as shown here (notice that we are no longer dealing with predicted values of Y, which are represented as $\hat{Y}_i$ in the equation above):

$$Y_i = a + bX_i + e_i$$

In words, what this equation says is that the observed value of the dependent variable Y_i is the sum of the intercept (a), the coefficient (or slope) (b) times the independent variable (X_i), plus some amount of error. The more variance in Y that X explains, the smaller the error (e_i) will be, on average. We will return to the concept of explained variance versus error in the next section.

7.2 Multiple Linear Regression

At this point, you are likely asking yourself what happened to the particularly useful ability of regression to account for multiple independent variables that predict a single dependent variable that I highlighted at the beginning of this chapter. Indeed, what has been presented so far is an improvement over correlation only insofar as we are interested in prediction and an easier substantive interpretation. Even

7.2 Multiple Linear Regression

then, this prediction is based on only one variable. To this point, we turn now to multiple linear regression. Multiple linear regression is where the utility of regression becomes more obvious, as the information that it provides is invaluable for responding to complex research questions. Because multiple linear regression can account for numerous independent variables in the same analysis, it is especially useful for individuals who conduct research in areas where randomized control trials (which isolate the impact of a particular variable through randomization) are not always possible (for more on randomized control trials, see Chap. 9). It can also be used for descriptive analysis when we want to account for the relationships between several independent variables and a particular dependent variable. Multiple linear regression helps us to partial out the relationship between an independent variable and a dependent variable while holding constant (controlling for) other variables that also serve to explain the dependent variable.

To illustrate multiple linear regression, we return to our working example of OLS regression and add a second predictor variable. In this case, in addition to using expenditure on instruction to explain international student enrollment, I add information about the per capita gross domestic product (GDP) of the state where an institution is located to our regression model. In equation form, this regression model would look something like this:

$$\hat{Y}_i = a + b_1 X_{1i} + b_2 X_{2i},$$

where, $\hat{Y}_i$ is a predicted number of enrolled international students, just like before, and a is the intercept of the regression model. X_{1i} refers to expenditure on instruction and X_{2i} is GDP, while b_1 is the coefficient associated with expenditure on instruction and b_2 is the coefficient associated with GDP. In this running example, this latter variable (GDP) is measured in units of US$1000. GDP is a logical predictor variable to enter into this regression model to the extent that we think that international students want to study in economically prosperous locations. A simplified form of the results of my regression analysis are found in Table 7.2 (more information later on why these results are simplified).

We interpret the coefficients found in the second column of Table 7.2 almost identically to the way we interpreted coefficients in a simple linear regression. The

Table 7.2 Multiple regression model predicting international student enrollment (simplified)

Variable	Coefficient
(Intercept)	66.82
Expenditure on Instruction (in millions)	6.15
GDP (in thousands)	1.24
N observations	50

Note Data source: Sample Dataset #1 - US National Center for Education Statistics, Integrated Postsecondary Education Data System

only difference is that we now interpret a coefficient while holding all other predictors in the model constant. In the case of GDP, we see that a one unit increase in a state's GDP (US$1000) positively relates to an increase in enrollment of one international student at that state's flagship institution *holding instructional expenditure constant*. This means that the coefficient for GDP is the relationship between GDP and international student enrollment after we account for the relationship between expenditure on instruction and international student enrollment. Notice, that the coefficient on expenditure on instruction is different from the one we estimated in the previous section. That is, a one unit increase in expenditure on instruction (US$1,000,000) is now related to an increase in enrollment of approximately 6 international students *holding state per capita GDP constant*. What the model is telling us is that if we had two institutions with the same value for expenditure on instruction, but one is located in a state with higher GDP than the other, we would expect the institutions to differ in the number of international students enrolled. To explore why that is, we take a quick excursion into what it means to *partial out* the variance that an independent variable explains.

7.2.1 Partialling Out

The concept behind partialling out the variance explained by independent variables in a multiple regression model is shown visually in Fig. 7.5 (note that another way of saying that we partial out the variance explained by independent variables is to say that these variables are *controlled for* or *held constant* in our regression model). In this figure, our dependent variable, international students, is depicted in light grey while our independent variables, expenditure on instruction and GDP, are depicted in darker shades of grey. The proportion of variance in international student enrollment that expenditure on instruction hypothetically explains is represented in the space where these two circles overlap in the top center of Fig. 7.5. Likewise, the proportion of the variance in international student enrollment that GDP explains is represented in the space where those two circles overlap in the middle left of the figure. Both these variables explain a unique portion of the variance in international student enrollment—represented in the area where the two circles, and only those two circles, overlap. A multiple linear regression model accounts for the unique portion of the variance explained by each independent variable. For this reason, the coefficient on instructional expenditure in our regression model decreased from 7 to 6 when we added GDP as an independent variable. What our new, multiple linear regression model tells us about is the relationship between instructional expenditure and international student enrollment *once we account for the variance explained by GDP*; that is, once we have partialled out the variance in international student enrollment explained by GDP.

Notice that in Fig. 7.5, there is a portion of the variance in international student enrollment that is shared between the two independent variables. In this case, a multiple linear regression cannot attribute explained variance to one variable or another. In reality, most independent variables share some portion of the variance

7.2 Multiple Linear Regression

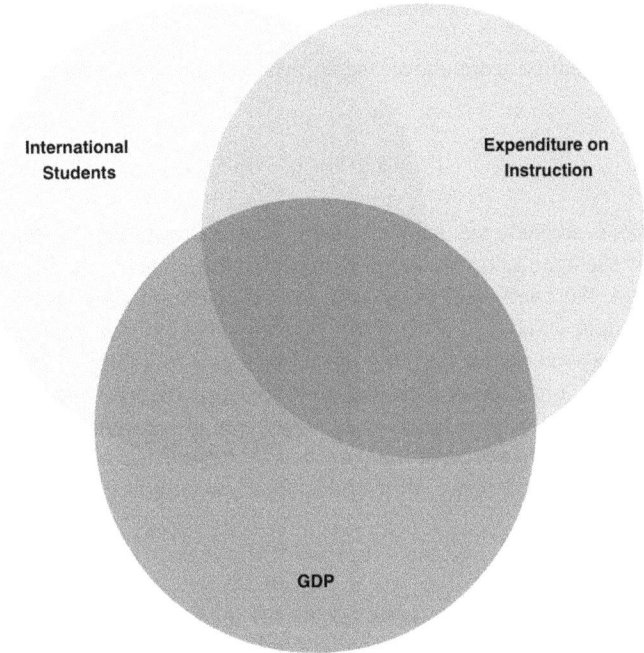

Fig. 7.5 Proportion of variance explained in multiple regression

in the dependent variable—that is, independent variables are often correlated with one another to a certain extent, and in this case, a regression model is unable to detect which variable is which. What is problematic, however, is if the shared variance overlaps too much. In these cases, a regression model is said to suffer from **multicollinearity** issues, and its output will not be especially meaningful. For example, if expenditure on instruction and GDP were correlated at 0.97, these two variables would be almost indistinguishable from one another in our regression model (for this reason, I double checked the correlation between the two variables and found that they were correlated at $r = 0.55$). It is useful to check correlations among predictor variables before you enter them into a multiple regression model to ensure that they do not highly correlate with one another. If you do find that two of your predictors are highly correlated, general wisdom is that you should enter only one of them into your regression model.[3]

---

[3] In other situations, independent variables can have what are called suppressor relationships with one another, meaning that the introduction of a given independent variable into a regression model can disguise the relationship between another independent variable and the outcome of interest. In the most extreme cases, introduction of a new variable into a regression model can change the sign (positive or negative) of the relationship between a given predictor and the outcome of interest. While a full discussion of suppressor relationships is beyond the scope of this book, this possibility

7.2.2 Full Function for Multiple Regression

To repeat the equation from above, the regression model we have estimated looks like this:

$$\hat{Y}_i = a + b_1 X_{1i} + b_2 X_{2i}$$

Again, $\hat{Y}_i$ is a predicted number of enrolled international students, a is the intercept of the regression model, X_{1i} refers to expenditure on instruction, X_{2i} is GDP, b_1 is the coefficient associated with expenditure on instruction, and b_2 is the coefficient associated with GDP. We can substitute the numbers from our estimated regression model into this equation, found in Table 7.2 ($\hat{Y} = 66.82 + 6.15X_1 + 1.24X_2$.) just like we did before to make predictions about international student enrollment, this time based on two variables. If we wanted to represent this regression model using observed values rather than predicted values, we would use an equation like the following, which adds the error term at the end:

$$Y_i = a + b_1 X_{1i} + b_2 X_{2i} + e_i$$

Notice that this is the exact same equation I used to represent a simple linear regression model, only there are now two independent variables instead of just one.

7.2.3 Hypothesis Testing for Regression Coefficients

For the remainder of this section, I turn to the reason why I referred to the regression model summarized in Table 7.2 as simplified. This table was missing two key pieces of information that we need to make sense of what a regression model tells us: significance tests of the model's coefficients and a measure of model fit.

A significance test for the coefficients in an OLS regression model is fairly straightforward. Just like we did in Chap. 6 when we conducted a significance test for a correlation coefficient, we perform a t-test for the coefficients in a regression model to explore whether they are significantly different from zero.[4] The null and alternative hypotheses that guide this test for significance are:

H_0 : The coefficient representing the relationship between the independent and the dependent variables is zero.

---

underscores the importance of carefully considering what independent variables should be included in a given statistical model.

[4] If pertinent to your research, it is possible to test other hypotheses about regression coefficients (i.e., that the coefficient is equal to a number aside from zero). The test of difference from zero, though, is the most common and is what statistical software report by default.

H_A : The coefficient representing the relationship between the independent and the dependent variables is not zero.

Notice that this null hypothesis is a two-tailed one—a relationship between the independent and dependent variables could be positive or negative. In equation form, we calculate the value of t as follows:

$$t = \frac{b - 0}{s_b}$$

Here, b is the coefficient from our regression model and s_b is its estimated standard error. We do not calculate s_b by hand in this book since it is a complex calculation and statistical software programs will do this calculation for you (it is a standard part of the output from a regression model in most statistical software programs). Once we arrive at a calculated value for t, we can find its associated p-value (derived, of course, from a t-table like the one in Appendix A; usually, your software program will perform this step for you, too) and make a determination about whether we reject or fail to reject the null hypothesis, just like we have done previously.

Table 7.3 adds information to the regression model summarized in Table 7.2 to include these hypothesis tests for the regression coefficients in our running example. This table is similar to what one might see in a published academic article, but sometimes the column of t-values is omitted. Notice that each coefficient is now accompanied by its corresponding standard error, in parentheses, and p-values for each hypothesis test are indicated with star levels. Information about how to interpret these two pieces of information is now included in a note at the bottom of the table.

To summarize this regression model in writing, we can say something like the following: "Both expenditure on instruction and a state's GDP predicted international student enrollment at standard levels of significance. Specifically, holding

Table 7.3 Multiple Regression Model Predicting International Student Enrollment (With Coefficient Hypothesis Testing)

Variable	Coefficient	T-value
(Intercept)	66.82	0.18
	(382.29)	
Expenditure on Instruction (in millions)	6.15***	7.19
	(0.86)	
GDP (in thousands)	1.24*	2.31
	(0.54)	
N observations	50	

Note Standard error in parentheses. ***$p < 0.001$, *$p < 0.05$ (two-tailed p-values). Data source: Sample Dataset #1 - US National Center for Education Statistics, Integrated Postsecondary Education Data System

GDP constant, a US$1,000,000 increase in expenditure on instruction was related to an average increase of approximately six international students ($p < 0.001$). While controlling for expenditure on instruction, a US$1000 increase in a state's GDP was associated with approximately one additional enrolled international student ($p < 0.05$)."

7.2.4 Model Fit

A final bit of information that we need to know about our regression model is an overall measure of how well the model predicts our outcome of interest, international student enrollment. The measure of model fit most commonly used for an OLS regression is called the **R-squared** (represented as R^2). Like the coefficient of determination in a correlation analysis (see Chap. 6), R^2 tells us the percentage of variance in the dependent variable that is explained by the independent variables in our regression model. In Table 7.4, I summarize both the regression models that we have estimated in this chapter—the simple linear regression model using only instructional expenditure as a predictor variable and the multiple linear regression model that adds GDP as a predictor. In addition to including significance tests for the coefficients in these models, I also include the calculated R^2 value for each model in the next to last row of the table.

Table 7.4 indicates that our simple linear regression explains around 66% of the variance in international student enrollment. This finding is likely because international students comprise a portion of total enrollment at a given institution, and institutions that enroll larger numbers of students overall likely also spend more on instruction. We should not read too much more into the relationship depicted

Table 7.4 Linear regression models predicting international student enrollment

Variable	Simple Linear Regression	Multiple Linear Regression
(Intercept)	16.73	66.82
	(396.51)	(382.29)
Expenditure on Instruction (in millions)	7.24***	6.15***
	(0.75)	(0.86)
GDP (in thousands)		1.24*
		(0.54)
N observations	50	50
R^2	0.66	0.70
Adjusted R^2	0.66	0.68

Note Standard errors in parentheses. ***$p < 0.001$, *$p < 0.05$ (two-tailed p-values). Data source: Sample Dataset #1 - US National Center for Education Statistics, Integrated Postsecondary Education Data System

in this regression. The multiple linear regression model explains about 70% of the variance in international student enrollment. This increase in R^2 compared to the simple linear regression makes sense—we allow for a second variable to explain some of the variance in international student enrollment, thus capturing some of the variance that was not captured in the simpler model. In practice, the more independent variables we add to our regression model, the greater the R^2 will be since each variable explains at least a small proportion of the variance in the independent variable. If we are really focused on explaining the most variance in our dependent variable as possible, we have a motivation to add as many independent variables as we can to the regression model. Unfortunately, this tactic would lead to some regression models that do not make much sense (in fact, this can be another version of the overfitting problem from Fig. 7.3). For example, adding a variable for the number of students at an institution that prefer to take notes using a pencil rather than a pen would increase our R^2 in this example even though we have no motivation for believing that this variable explains international student enrollment at all. For this reason, it is important to select the variables that go into a regression model based on a theoretical or conceptual framework that guides your research and, at the very least, common sense.

Statistically, we can also penalize a regression model's calculated R^2 for the number of independent variables that it contains, thus providing a more conservative measure of model fit. Notice that in the last row of Table 7.4, I include an adjusted R^2 value. This measure of model fit takes into account the number of predictors entered into a regression model and is typically lower (although sometimes not much lower) than the calculated R^2 value. For this reason, many researchers prefer an adjusted R^2 to the true R^2 when reporting the results of a regression model. While it is possible for the adjusted R^2 to shrink when more predictors are added to the model, this is not the case in our running example. Even the adjusted R^2 shows some improvement (from 0.66 to 0.68) when we add per capita GDP as a predictor in our regression model.

7.3 Example from the Literature

To explore an example of ordinary least squares in action, this section draws from Schmidt (2020), a study that used OLS regression to explore the relationship between an institution of higher education's expenditures on instruction, research, public service, academic support, student services, and institutional support on the one hand and an institution's level of undergraduate international student success, defined as the six-year baccalaureate degree completion rate, on the other. In brief, this study was motivated by the fact that international students often pay higher tuition and fee rates at public higher education institutions in the United States and, thus, it is important to understand if and how students benefit from the way that institutions spend these additional funds. The data that inform this study come from the Integrated Postsecondary Education Data System (IPEDS) in the United States and represent 414 public doctoral and master's level colleges and

Table 7.5 OLS regression model estimating the relationship between institutional expenditures and undergraduate international student graduation rate, intercept and control variables omitted (Schmidt, 2020, pp. 656–657)

Variable	Coefficient
Instructional expenditure	0.078
	(0.048)
Research expenditure	0.000
	(0.037)
Public service expenditure	−0.110
	(0.076)
Academic support expenditure	0.217*
	(0.108)
Student services expenditure	0.002
	(0.130)
Institutional support expenditure	−0.042
	(0.110)
R^2	0.321
N	414

Note Robust standard errors in parentheses. *$p < 0.05$

universities. To account for fluctuations in variables corresponding to institutional expenditures and international student success, Schmidt (2020) averages data corresponding to cohorts of international students entering institutions in 2009, 2010, and 2011 to arrive at a dataset that represents three-year mean values for each of the variables in this study.

The outcome of interest in this study was the institution's six-year international student graduation rate, as averaged over the three cohorts of students. The predictors of interest were the expenditure variables listed in the previous paragraph, such as expenditures on instruction and academic support. The study's most robust regression model, reproduced here in Table 7.5, is an OLS regression model that takes the following general form:

$$GradRate_i = a + Expenditures_i b_1 + Controls_i b_2 + e_i$$

Here, $GradRate_i$ represents the dependent variable (each institution's average six-year graduation rate for international students), a is the intercept, b_1 represents a vector of coefficients corresponding to the predictors of interest (various categories of expenditure) and b_2 corresponds to coefficients for a group of additional independent variables that we will refer to here as control variables.[5] Since these control variables contain some complexities that have not yet been introduced in

---

[5] This group includes an institution's Carnegie classification, size, percentage of enrolled students classified as international, and out-of-state tuition and fee charges. Note that because both b_1 and b_2 represent coefficients corresponding to multiple variables, the convention is to reverse the order of the coefficients and variable names in the regression equation.

this book and that are not central to our discussion here, we will ignore them for the time being. For simplicity's sake, I have left them out of Table 7.5, but note that they are still included in the regression model that produced the results in this table. The fact that these variables are present in the statistical model is important, as they represent other factors that likely come into play when considering international student graduation rate.

An important caveat for interpretation in this regression model is that all expenditure variables are represented in 100s. That is, a one-unit increase in all these variables represents $100. The results in Table 7.5 suggest that, among the expenditure predictors of interest, only expenditure on academic support has a significant relationship with international student graduation rate, as indicated by the p-value corresponding to this variable's coefficient. More specifically, a $100 increase in expenditure on academic support is related to a 0.217 percentage point increase in an institution's six-year international student graduation rate. These results suggest that academic support may be key to international student success. Schmidt (2020) would argue that institutional expenditure in this category is fair to the extent that international students typically pay more for their educations than do domestic students. The R^2 found at the bottom of Table 7.5 (Schmidt does not provide an adjusted R^2 value) suggests that the independent variables in this regression model (both the predictors of interest and the control variables not included in Table 7.5) explain around 32% of the variance in international student graduation rates.

7.4 Another Note on Correlation and Causation

Before ending this chapter, I return to a point that I have already made several times throughout this book—that correlation is not causation. While regression analysis, and particularly multiple regression analysis, is a powerful statistical technique with many advantages, these statistical models remain correlational, or descriptive, rather than causal without the proper research design. This can be easy to forget because of the asymmetric interpretation of regression coefficients. In other words, in the example we just outlined, just because there is a significant relationship between academic support expenditures and an institution's international student graduation rate, does not mean that an increase in expenditure on academic support *causes* international students to perform better academically. While the next chapter expands on the regression concepts introduced in this chapter, Chap. 9 (Introduction to Experimental and Quasi-experimental Design) takes up the notion of causality again in a discussion of how we can use OLS regression with the proper research design to approach causal interpretations.

7.5 Practice Problems

1. Sample Dataset #2 contains information from 161 U.S. Liberal Arts institutions from the 2016–17 academic year. Here, we focus on the number of students who studied abroad from these institutions (represented by the variable study abroad). The table below summarizes an OLS regression model that takes the following form:

$$TotalSA_i = a + b_1 TotalEnroll1000_i + b_2 PctFemale_i + b_3 AcceptRate_i + e_i$$

This regression uses a liberal arts institution's total enrollment (measured in 1000s) (totalenroll), percentage of students identifying as female (pctfemale), and acceptance rate (acceptrate) to predict total study abroad participation.

Variable	Coefficient
(Intercept)	259.38***
	(48.20)
Total enrollment (in 1000s)	36.36***
	(7.45)
Percent female	−0.60
	(0.65)
Acceptance rate	−2.12***
	(0.44)
N observations	161
R^2	0.23
Adjusted R^2	0.22

Note Standard errors in parentheses. ***$p < 0.001$. Data source:Sample Dataset #2 - US National Center for Education Statistics, Integrated Postsecondary Education Data System

a. Interpret the coefficients corresponding to total enrollment, percent female, and acceptance rate in this regression model. Which ones are statistically significant?
b. What percentage of the variance in study abroad participation does this regression model explain?
c. Write out the equation that corresponds to this regression model. What is the predicted number of students who study abroad at liberal arts institutions with the following characteristics:
   i. Total Enrollment = 7000, Percent female = 67%,[6] Acceptance rate = 20%
   ii. Total Enrollment = 300, Percent female = 55%, Acceptance rate = 10%

---

[6] Note that numbers for percent female and acceptance rate should be taken as percentages (67%) rather than proportions (0.67) for the purposes of this practice problem.

iii. Total Enrollment = 750, Percent female = 98%, Acceptance rate = 5%
2. Now use your statistical software program of choice to load Sample Dataset #2 and run this regression model yourself (note that you will have to create the enrollment in 1000s variable yourself). Double-check your work to ensure that you arrived at the same numbers as in the table above. In a second step, add two additional predictor variables to the model: the percentage of the student body that is comprised of graduate students (pctgrad) and the percentage of the student body receiving Pell grants (pctpell). The equation version of this model is as follows:

$$TotalSA_i = a + b_1 TotalEnroll1000_i + b_2 PctFemale_i + b_3 AcceptRate_i \\ + b_4 PctGrad_i + b_5 PctPell_i + e_i$$

a. Interpret the coefficients corresponding to total enrollment, percent female, acceptance rate, percent graduate student, and percent Pell recipient in this regression model. Which ones are statistically significant?
b. Are your results for total enrollment, percent female, and acceptance rate different from the first model you ran? Why?
c. Does this second model explain more or less of the variance in study abroad participation compared to the first model? Why? How does the penalized fit from the adjusted R^2 compare?
d. If you were the Director of Study Abroad at a liberal arts institution, how could you use the results of this regression model in your work?

Recommended Reading

A Deeper Dive

Gujarati, D. N., & Porter, D. C. (2010). Basic ideas of linear regression: The two-variable model. In *Essentials of econometrics* (pp. 21–52). McGraw-Hill.
Gujarati, D. N., & Porter, D.C. (2010). Multiple regression: Estimation and hypothesis testing. In *Essentials of econometrics* (pp. 93–131). McGraw-Hill.
Urdan, T.C. (2017). Regression. In *Statistics in plain English* (pp. 183–204). Routledge.
Wheelan, C. (2013). Regression analysis: The miracle elixir. *Naked Statistics: Stripping the Dread from Data* (pp. 185–211). Norton.
Wheelan, C. (2013). Common regression mistakes: The mandatory warning label. In *Naked Statistics: Stripping the Dread from Data* (pp. 212–224). Norton.
Wooldridge, J. M. (2016). The simple regression model. In *Introductory econometrics: A modern approach* (pp. 20–59). Cengage Learning.
Wooldridge, J. M. (2016). Multiple regression analysis: Estimation. In *Introductory econometrics: A modern approach* (pp. 60–104). Cengage Learning.
Wooldridge, J. M. (2016). Multiple regression analysis: Inference. In *Introductory econometrics: A modern approach* (pp. 105–148). Cengage Learning.

Additional Examples

Kritz, M. M. (2016). Why do countries differ in their rates of outbound student mobility? *Journal of Studies in International Education, 20*(2), 99–117.

Schmidt, A. (2020). Are international students getting a bang for their Buck?: The relationship between expenditures and international student graduation rates. *Journal of International Students, 10*(3), 646–663.

Whatley, M., Landon, A. C., Tarrant, M. A., & Rubin, D. (2021). Program design and the development of students' global perspectives in faculty-led short-term study abroad. *Journal of Studies in International Education, 25*(3), 301–318.

Additional Regression Topics

8

The previous chapter introduced key concepts in ordinary least squares (OLS) regression, including functional form, simple and multiple linear regression, and hypothesis testing for regression coefficients. While the information in Chap. 7 is foundational for more advanced regression models, it is the beginning rather than the end of what regression can do. The purpose of this chapter is to introduce additional topics in regression that are especially useful in international education research. While the topics covered here are certainly not a complete list of everything that regression analysis can do, they represent a starting point for thinking about regression models in different ways.

This chapter is divided into four parts. The first section of this chapter considers other variable types that can be included as predictors in regression models. You may have noticed in the previous chapter that all the predictor variables that we focused on were continuous, such as GDP and expenditure on instruction. Of course, in real life, categorical variables also serve to explain outcome variables. For example, in our running example of an OLS regression model predicting international student enrollment, certain descriptive information about the institution in question may be important to consider, such as an institution's geographic location. Geographic location is not a continuous variable but rather a categorical one. We consider these kinds of variables in the first section of this chapter.

The second section of this chapter discusses another variable type that can be included as a predictor in a regression model—interaction terms (also referred to as simply 'interactions'). We might hypothesize that estimated the relationship between two variables in a regression model, such as expenditure on instruction and international student enrollment, is different depending on the value of another variable, such as whether the institution operates a hospital, which might attract international students interested in medicine or health care, who also care about the quality of the instruction they receive. The relationship between expenditure on instruction and international student enrollment may be different at institutions that have a hospital compared to those that do not. That is, in more technical terms, we hypothesize that there is a significant interaction effect between instructional expenditure and a binary indicator of having a hospital on campus in a regression model predicting international student enrollment.

The third part of this chapter provides a more in-depth consideration of functional form. We introduced functional form in Chap. 7, noting that while many options exist for a regression model's functional form, our running example considered only a straight line. Now, we introduce a quadratic functional form, which is more complex. This kind of functional form is important when the relationship between two variables is u-shaped, such as in Fig. 7.2.

The fourth section of this chapter introduces regression models for binary outcome variables, and specifically logistic regression. Many of the outcome variables that might be of interest to international educators are binary, such as whether a student studies abroad or whether a faculty member conducts collaborative international research. Unlike the outcome variables that we have considered thus far, these variables cannot be plotted continuously along a vertical axis, as they only take two values (yes and no, for example). This situation requires consideration of a different kind of regression analysis.

8.1 Categorical Predictors

Researchers use what are called **dummy variables** to encode categorical (i.e., non-continuous) variables so that they can be entered into regression models (some researchers call these **qualitative variables**). Chapter 1 introduced two other names that can also be used to describe this kind of variable, namely discrete and dichotomous variables. Examples of categorical variables that might be important in a dataset of people include an individual's racial/ethnic or gender identity, the religion that they practice, or their nationality. When we think about institutions, categorical variables include characteristics like geographic location or types of educational offerings. Categorical variables that pertain to countries might include world region or type of government. In this book, we have already seen two types of analyses that include categorical variables as related to a continuous variable. We introduced *t-tests* in Chap. 4, which are used when a categorical variable is binary (has only two categories). In Chap. 5, we saw ANOVA analysis, which is used with a categorical variable that has more than two categories. Using categorical variables in a regression context extends these two analytic approaches. Indeed, in an OLS regression analysis that includes a single categorical variable as a predictor, regression yields the same substantive results as these simpler analyses.

When a categorical variable has only two categories, entering it into a regression analysis is relatively straightforward. Drawing on the running example in Chap. 7, we can expand our regression model predicting international student enrollment at US flagship institutions (y_i) to include a variable indicating whether an institution operates a hospital on campus. International students, especially those wanting to enter a health profession, might be attracted to the opportunity for hands-on

8.1 Categorical Predictors

experience on the same campus where they study. In this situation, our regression equation now looks like this:

$$Y_i = a + b_1 X_{1i} + b_2 X_{2i} + b_3 X_{3i} + e_i$$

where X_1 is expenditure on instruction and X_2 is GDP, just like before, and X_3 is a binary indicator of whether an institution operates an on-campus hospital. For most software programs, this binary indicator has to be numerical in format, which requires us to transform data stored as text into numbers. For example, in this case, we might have a dataset that contains "yes" for institutions with hospitals and "no" for those without. For binary categorical variables, converting text into numbers typically means that we choose one category that is represented as a 1 (e.g., "yes") while the other is represented as a 0 (e.g., "no"). This conversion transforms the variable into what we refer to as a dummy variable and using 0 and 1 as our coding choices makes interpretation of the model considerably easier later. Specifically, we can interpret the regression coefficients that result from this analysis as if we were flipping on a light switch – that is, what is the predicted increase or decrease in the dependent variable when this dummy variable switches from off (0 = no hospital) to on (1 = hospital)?

The results of the regression analysis described in this equation are summarized in Table 8.1. Notice that the addition of another variable (Hospital) means that the coefficients for both expenditure on instruction and GDP shift around a bit if we compare them to the results in Table 7.4, but these numbers do not change substantially. Both coefficients also remain significant predictors of international enrollment at the $p < 0.05$ level or lower. The coefficient on Hospital, however, is not significant (the t value for this variable is quite low: $t = -0.36$). It seems

Table 8.1 OLS regression model predicting international student enrollment

Variable	Coefficient	T-value
(Intercept)	64.10	0.17
	(385.96)	
Expenditure on Instruction (in millions)	6.29***	6.65
	(0.95)	
GDP (in thousands)	1.21*	2.21
	(0.55)	
Hospital	−178.32	−0.36
	(495.54)	
N observations	50	
R^2	0.70	
Adjusted R^2	0.68	

Note Standard error in parentheses
*** $p < 0.001$, * $p < 0.05$ (two-tailed p-values)
Data source: Sample Dataset #1 - US National Center for Education Statistics, Integrated Postsecondary Education Data System

that having a hospital on campus does not significantly relate to the number of international students that enroll at a particular flagship institution. Even though it is not significant, we can still interpret this coefficient for illustrative purposes. In this example, the Hospital variable is coded 1 if an institution has an on-campus hospital and 0 if not. Here, I would say that flagship institutions that have a hospital on campus (the 1 category) are predicted to enroll around 178 fewer international students compared to flagship institutions that do not have a hospital on campus (the 0 category). The direction of the comparison here, that we are comparing campuses with hospitals to those that do not, is important. When you include a dummy variable in a regression model, the category coded with a zero (no hospital, in this case) is the **reference**, or **comparison, category**, meaning that results are always interpreted in relationship to that category.

We can expand this concept of a binary dummy variable to include variables that have multiple categories, similar to how we arrived at ANOVA in Chap. 5 by expanding on the *t*-test. Here, we add a categorical variable corresponding to the US region where a state flagship institution is located to our running example. We might suspect that international students find some regions more attractive than others, whether due to climate or politics or some other reason. The regions found in this dataset correspond to those of the US Bureau of Economic Analysis. These regional classifications are based on a series of economic, demographic, social, and cultural characteristics of the 50 US states and Washington, DC and include eight regions, listed in Table 8.2.

To enter these variables into a regression analysis, we first have to transform them into a series of dummy variables using 0's and 1's, just like we did when we had a variable with only two categories. To do this, we create new dummy variables for each region, which are then coded with a 1 if an institution is located in that region and a 0 if it is not. The result of this variable creation looks something like what is found in Table 8.3. In this table, Institutions 1 and 2 are located in states in the Southwest, and as a consequence receive 1's in the 'Southwest' column. Notice that these two institutions receive 0's for all other regions. The same is true of institution 3. This institution is located in the Plains region and receives a 1 only in the 'Plains' column. Important to note is that the categories for these

Table 8.2 US states in eight regions

Region	States
New England	CT, ME, MA, NH, RI, VT
Mid East	DE, DC, MD, NJ, NY, PA
Great Lakes	IL, IN, MI, OH, WI
Plains	IA, KS, MN, MO, NE, ND, SD
Southeast	AL, AR, FL, GA, KY, LA, MS, NC, SC, TN, VA, WV
Southwest	AZ, NM, OK, TX
Rocky Mountains	CO, ID, MT, UT, WY
Far West	AK, CA, HI, NV, OR, WY

8.1 Categorical Predictors

Table 8.3 Dummy coding example for a variable with more than two categories (US Region)

Institution	N England	Mid East	Great Lakes	Plains	South east	South west	Rocky Mountains	Far West
1	0	0	0	0	0	1	0	0
2	0	0	0	0	0	1	0	0
3	0	0	0	1	0	0	0	0
4	1	0	0	0	0	0	0	0
5	0	1	0	0	0	0	0	0
6	0	0	0	0	1	0	0	0
7	0	0	1	0	0	0	0	0
8	0	0	1	0	0	0	0	0
9	0	0	0	0	0	0	1	0
10	0	0	0	0	0	0	0	1
11	0	0	0	0	0	0	0	1

Region variables are mutually exclusive. An institution cannot be located in more than one region.

When adding dummy variables corresponding to a categorical variable with more than two categories in a regression model, you must choose a reference category for the purpose of interpreting your results. In fact, a regression model that includes all the categories is not possible since these categories exhibit perfect multicollinearity with one another, as they are mutually exclusive.[1] While sometimes the choice of which category to designate as the reference group is obvious (perhaps it is related to your theoretical framework or one of your research questions), other times, the choice of a reference category is arbitrary. In this example, I chose Far West as the reference category, and as a consequence, I interpret my results with reference to the Far West (e.g., "Compared to institutions in the Far West, institutions in X region were predicted to enroll a greater number of international students, on average"). I did not have any particular motivation for making this choice—I made Far West my reference category because I needed one for the purposes of both interpreting coefficients and simply getting my regression model to run.[2]

The regression model including these regional dummy variables is found in Table 8.3. We can write this regression model in equation format as follows:

$$Y_i = a + b_1 X_{1i} + b_2 X_{2i} + b_3 X_{3i} + X_{4i} b_4 + e_i$$

where all the terms are defined as before, and X_{4i} represents all eight of the regions in the dataset. b_4, then, is a vector of regression coefficients, one corresponding

---

[1] A more in-depth discussion of multicollinearity is found in Chap. 7.
[2] Note, however, that if you are interested in making region-specific predictions, you will arrive at the same predictions regardless of which region you choose as the reference group.

to each region (besides Far West, which is the reference group). If I wanted or needed to be more explicit about the variables in my regression model, I could write this equation in a longer format as follows:

$$Y_i = a + b_1X_{1i} + b_2X_{2i} + b_3X_{3i} + b_4 NewEngland$$
$$+ b_5 MidEast + b_6 GreatLakes$$
$$+ b_7 Plains + b_8 Southeast + b_9 Southwest + b_{10} RockyMtns + e_i$$

In this longer equation, it becomes clear that in this regression model, each region (besides Far West, which is the reference group) gets its own regression coefficient (e.g., b_4 refers only to institutions in New England). It is also clear that Far West is the reference group as it is explicitly left out of the equation. However, this longer notation can become very complicated very quickly if there are many variables with many categories in your regression model, so researchers often prefer the shorter, first equation provided here.

Notice first that in the regression model in Table 8.4, both expenditure on instruction and GDP remain statistically significant predictors of international student enrollment ($p<0.01$ or lower) even though their coefficient values shift a little bit with the addition of the regional predictor variables. Similarly, the variable corresponding to whether an institution has an on-campus hospital remains non-significant ($p>0.05$). Two regions, the Mid East and Great Lakes regions, are predicted to enroll significantly more international students, as evidenced by their positive coefficients ($p < 0.05$ and $p < 0.001$, respectively). The interpretation of these significant coefficients depends on the region that serves as the reference category—Far West (notice, too, that I note this reference category in the notes at the bottom of the table). That is, institutions in the Mid East region enrolled around 1,806 more international students *compared to institutions in the Far West* ($p < 0.05$) while institutions located in the Great Lakes region enrolled around 2,884 more international students, also *compared to institutions in the Far West* ($p < 0.001$).

8.2 Interactions Between Variables

Another kind of variable that you might want to include in a regression model is an **interaction variable**. The purpose of an interaction variable is to explore whether the relationship between a given predictor variable and the outcome variable changes depending on the value of another predictor variable. For example, we might hypothesize that the relationship between instructional expenditure and international student enrollment is different for institutions that have a hospital on campus compared to those that do not, as posited earlier in this chap.[3] To test

---

[3] Here, we interact a continuous variable (instructional expenditure) and a dummy variable (having a hospital). However, interactions can occur between all types of variables, including two dummy variables and two continuous variables.

8.2 Interactions Between Variables

Table 8.4 OLS regression model predicting international student enrollment

Variable	Coefficient	T-value
(Intercept)	230.90	0.38
	(612.98)	
Expenditure on Instruction (in millions)	4.50***	5.02
	(0.90)	
GDP (in thousands)	1.50**	3.22
	(0.47)	
Hospital	−239.23	−0.58
	(413.65)	
New England	−174.56	−0.25
	(696.59)	
Mid East	1806.05*	2.52
	(715.88)	
Great Lakes	2883.63***	3.88
	(743.75)	
Plains	1115.92	1.67
	(667.93)	
Southeast	−338.48	−0.58
	(588.64)	
Southwest	269.72	0.36
	(745.51)	
Rocky Mountains	203.41	0.28
	(735.57)	
N observations	50	
R^2	0.83	
Adjusted R^2	0.79	

Note Standard error in parentheses
*** $p < 0.001$, ** $p < 0.01$, * $p < 0.05$ (two-tailed *p*-values).
Reference category for region is Far West
Data source: Sample Dataset #1 - US National Center for Education Statistics, Integrated Postsecondary Education Data System

this hypothesis, we need to include an interaction term between our variable corresponding to instructional expenditure and our variable corresponding to having a hospital in our regression model. To create an interaction term, we simply multiply the two variables together (researchers often do this variable creation in the data cleaning process when they are preparing their data for analysis, but many statistical software programs will also allow researchers to multiply two variables directly in a regression analysis). In this case, since our Hospital variable is coded 1 for institutions that have a hospital on campus and 0 otherwise, we end up with a variable that accounts for instructional expenditure only at institutions with a hospital (that is, the instructional expenditure amount for institutions without a hospital gets multiplied by 0). The regression model with this interaction term in equation form is as follows:

$$Y = a + b_1 X_{1i} + b_2 X_{2i} + b_3 X_{3i} + X_{4i} b_4 + b_5 X_{1i} * X_{3i} + e_i$$

In this equation, all terms are defined as we have been defining them. That is, X_{1i} is instructional expenditure, X_{2i} is GDP, X_{3i} is having a hospital, and X_{4i} corresponds to our vector of region dummies. The final term in this regression model (not counting the error term) is $X_{1i} * X_{3i}$, instructional expenditure multiplied by the hospital dummy variable. When you enter an interaction term into your regression model, you should also enter in main effects for the original variables that you used to create the interaction (some researchers refer to these as conditional effects because of the way they are interpreted, as described in the next paragraph). In practice, this means that you enter the two interaction term variables by themselves (the main effects) and the variable that multiplies them (the interaction term) into the regression model at the same time. This practice ensures that any relationship between the interaction term and the outcome variable is *above and beyond* the primary relationship between the two variables and the outcome variable.

When I ran this regression model, I found the results in Table 8.5. In this model, the interaction term between instructional expenditure and having a hospital is significant ($p < 0.05$) while the main effect for instructional expenditure remains significant (p < 0.001) and the main effect for hospital remains non-significant ($p > 0.05$), similar to the model in Table 8.4, which did not include the interaction term. Because the interaction term is significant, we can conclude that it is probably important in predicting international student enrollment. In other words, the effect of instructional expenditure on international student enrollment likely depends on whether the institution has a hospital or not. Once we add an interaction term to a model, the main effects are interpreted as a predictor's effect when the conditioning variable is equal to 0 (for example, the effect of instructional expenditure when an institution does not have a hospital, i.e., Hospital = 0) and all other variables are held equal. This is why some researchers refer to main effects as conditional effects. To interpret these specific regression results, we can say that instructional expenditure has a positive relationship with international student enrollment at institutions without a hospital, as suggested by the positive coefficient on instructional expenditure ($b_1 = 6.23$, $p < 0.001$), but that this positive relationship is attenuated somewhat at institutions that have a hospital on campus, as suggested by the negative coefficient on the interaction term ($b_5 = -3.11$, $p < 0.05$).[4]

---

[4] With a bit of extra work, a model like this allows the researcher to determine if the effect of instructional expenditure remains positive and significant for institutions with a hospital. Interested readers are encouraged to look up *post hoc t-ratios* to see how to do this.

8.2 Interactions Between Variables

Table 8.5 OLS regression model predicting international student enrollment

Variable	Coefficient	T-value
(Intercept)	395.92	0.79
	(504.56)	
Expenditure on Instruction (in millions)	6.23***	4.96
	(1.26)	
GDP (in thousands)	0.95	1.93
	(0.49)	
Hospital	1382.66	1.67
	(828.34)	
New England	−681.05	−1.19
	(571.16)	
Mid East	1292.41*	2.09
	(618.19)	
Great Lakes	2246.11**	3.39
	(662.92)	
Southeast	−1169.44*	−2.52
	(463.59)	
Southwest	−599.38	−0.90
	(663.75)	
Rocky Mountains	−409.35	−0.67
	(608.53)	
Expenditure on Instruction (in millions) *Hospital	−3.11*	−2.12
	(1.47)	
N observations	50	
R^2	0.84	
Adjusted R^2	0.80	

Note Standard error in parentheses
*** $p < 0.001$, ** $p < 0.01$, * $p < 0.05$ (two-tailed *p*-values). Reference category for region is Far West
Data source: Sample Dataset #1 - US National Center for Education Statistics, Integrated Postsecondary Education Data System

An often more intuitive way to interpret a significant interaction effect is to plot predicted values from the interaction so that you can visually explore the relationships among variables. Figure 8.1 plots the predicted relationship between instructional expenditure as conditioned by the presence or absence of a hospital and international student enrollment from the regression model in Table 8.5. This plot contains two lines, one for institutions that have a hospital (the grey line) and another for institutions that do not (the black line). Amounts of instructional expenditure are found along the horizontal axis of Fig. 8.1 while the number of

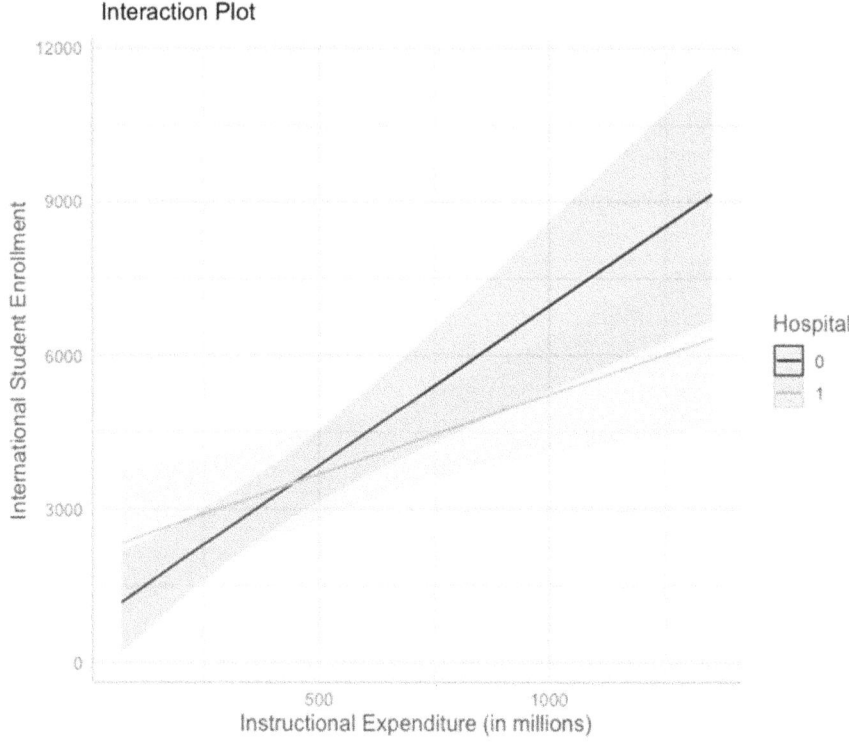

Fig. 8.1 Plot of the effect of the interaction between instructional expenditure and having a hospital on international student enrollment (*Data source:* Sample Dataset #1 - US National Center for Education Statistics, Integrated Postsecondary Education Data System)

international students that the regression model (Table 8.5) predicts to enroll at these institutions (the outcome variable) is on the vertical axis. As this figure clearly demonstrates, instructional expenditure has a positive relationship with international student enrollment regardless of whether the institution has a hospital. However, the slope of the line corresponding to institutions that do not have a hospital is steeper than the slope of the line corresponding to institutions that do. That is, a one-unit increase in instructional expenditures (which is scaled in millions) at an institution without a hospital results in a greater predicted increase in international student enrollment compared to a one-unit increase in instructional expenditure at an institution with a hospital.[5]

---

[5] When drawing a graph like this one, the other control variables should be set to a central value, such as the mean or mode. This ensures that predicted values are realistic.

8.3 Functional Form

Creating interaction terms to enter into linear regression models is one way that you might manipulate variables to explore more complex relationships between variables in a dataset. Another way to explore more complex relationships among variables is to consider non-linear functional forms, meaning that the relationship between a predictor variable and an outcome variable is not a straight line. In more technical terms, the slope of the regression line is not constant for all values of a predictor variable. We briefly mentioned functional form when discussing Fig. 7.2 in the previous chapter, suggesting that a u-shape of some sort might be a better functional form for describing the relationship between instructional expenditure and international student enrollment. We have been assuming that a straight line best depicts the relationship between these two variables, but this may not be the case. Using the incorrect functional form can have important consequences for the validity of a regression model, and the incorrect functional form can lead to biased results.

One common **non-linear functional form** (here, we say 'non-linear' to mean 'not-a-straight-line') in education research is the quadratic functional form. Figure 8.2 illustrates this functional form along with, for comparison purposes, the linear functional form that we have been assuming all along for the relationship between international student enrollment and instructional expenditure. Notice that the quadratic functional form is slightly curved compared to the straight line represented in the linear functional form.

At a glance, the quadratic functional form appears to fit the data from our running example better than the linear functional form we have been using. But how can we choose between the two forms in a way that is principled? The answer

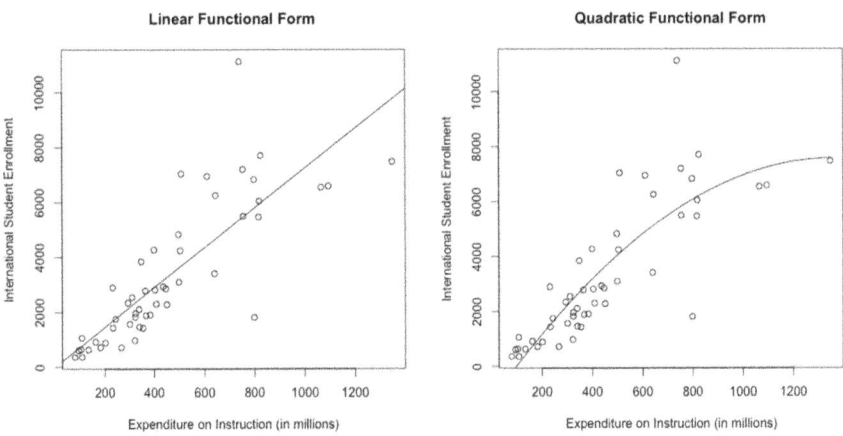

Fig. 8.2 Functional form options depicting the relationship between international student enrollment and expenditure on instruction (*Data source:* Sample Dataset #1 - US National Center for Education Statistics, Integrated Postsecondary Education Data System)

to this question lies in the residuals of the regression models using each of these alternative functional forms.

Remember that in Chap. 7, we defined the residuals of the regression model as the variance in the dependent variable that gets "left over" when we run a regression model. Another way of saying this is that the residuals of a regression model are represented in the space between the data points and the regression line itself. The goal of regression modeling is to minimize this space. In an ordinary least squares regression model, we assume that the residuals, or error, are random, meaning that we have included all the independent variables in our regression model that really matter for predicting the dependent variable. What is left over—the residuals—is just random noise. When we look at the residuals, we hope that they hover around a mean value of zero and display no clear pattern.

To illustrate this point, we return to the first, simple linear regression model that we explored in Chap. 7. This model takes the following form, and assumes a linear relationship between international student enrollment (Y_i) and instructional expenditure (X_i):

$$Y_i = a + bX_i + e_i.$$

The results of this model are summarized again for you in the first column in Table 8.6. An important point to make here is that while e_i in a regression model can feel like a catch-all term in a regression equation, meaning that it captures all the variation in Y_i that the other variables do not, we can actually calculate e_i for each prediction in a dataset. That is, the distance between the regression line and each real observation in a dataset is information that is knowable. Once we calculate the error—or residual—for each observation in the dataset, we can plot the residuals in a visual display that looks something like Fig. 8.3, which graphs

Table 8.6 Ordinary least squares regression models predicting international student enrollment using linear and quadratic functional forms

Variable	Linear	Quadratic
(Intercept)	16.73	−1214.50
	(398.51)	(626.51)
Expenditure on instruction (in millions)	7.24***	12.91***
	(0.75)	(2.41)
(Expenditure on Instruction)²		−0.01*
		(0.00)
N observations	50	50
R^2	0.66	0.70
Adjusted R^2	0.66	0.69

Note Standard errors in parentheses
*** $p < 0.001$, * $p < 0.05$ (two-tailed p-values)
Data source: Sample Dataset #1 - US National Center for Education Statistics, Integrated Postsecondary Education Data System

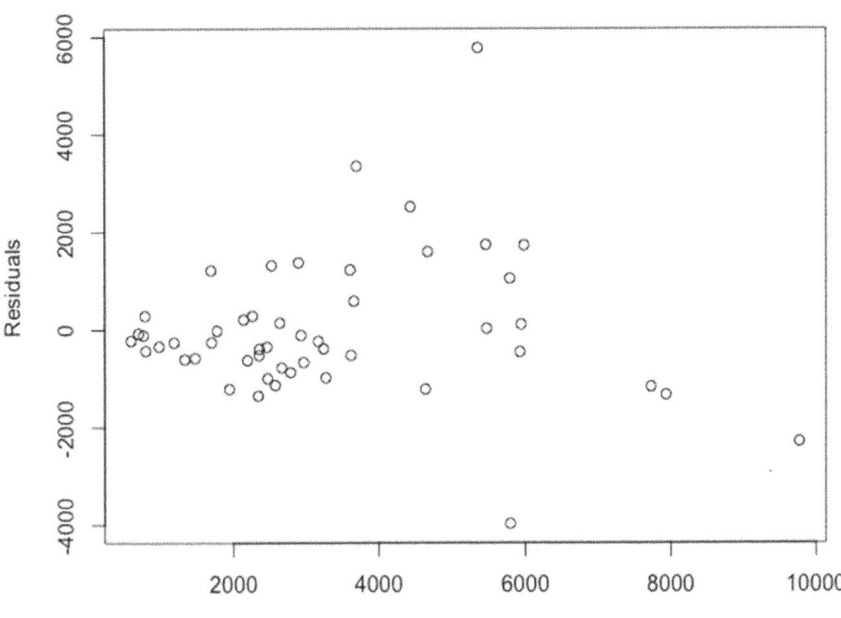

Fig. 8.3 Residuals from a linear functional form OLS regression model using instructional expenditure to predict international student enrollment (*Data source:* Sample Dataset #1 - US National Center for Education Statistics, Integrated Postsecondary Education Data System)

the residuals from the simple linear regression model in this equation.

The plot in Fig. 8.3 suggests first that the residuals of this regression model do not cluster very closely around zero (notice that one prediction is off by almost 6000 students!). Moreover, at high predicted values, we consistently get negative residuals, and we would like the residuals to hover around zero for all predicted values, even particularly high ones. A linear functional form may not be the best fit for our data.

8.3.1 Quadratic Functional Form

The **quadratic functional form** is one among many functional forms that we might try to fit to our data and is a common functional form that you will see in education research. Generally speaking, a quadratic functional form fits data in a u- or inverse-u-shape, as illustrated in Fig. 8.4 (in the left and right panels of this figure, respectively). In practice, this functional form indicates a negative (or positive) relationship between the independent and dependent variables up to

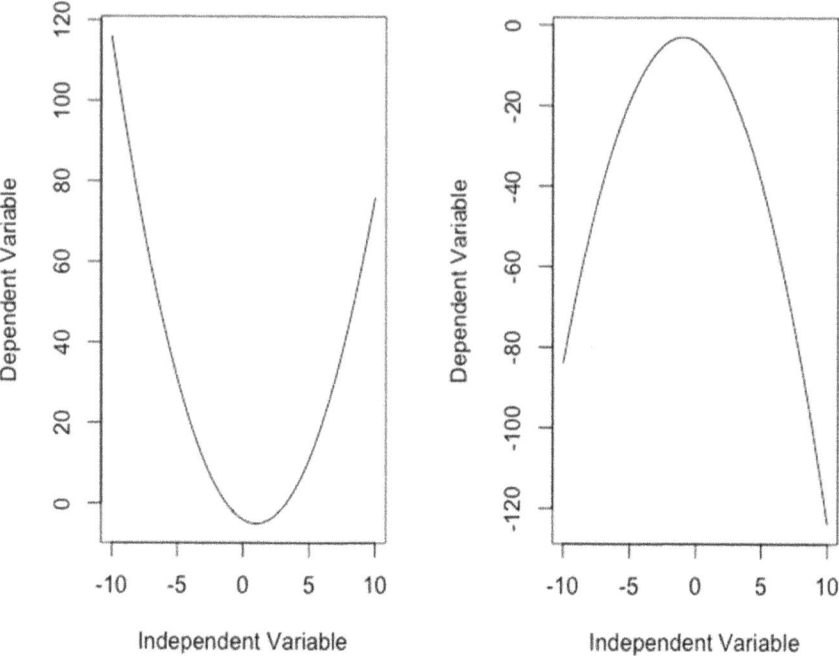

Fig. 8.4 Quadratic functional form examples (*Note* Data in this figure are invented for illustrative purposes.)

a certain value of the independent variable, where the relationship changes direction, exhibiting a positive (or negative) relationship afterwards. In our running example, we might think that at low levels of expenditure on instruction, adding additional funds entices more international students to enroll. However, once an institution reaches a certain point of expenditure on instruction, this positive relationship may start to subside and eventually give way to a negative one, wherein adding more funds to instructional expenditure actually deters international students from enrolling (e.g., perhaps because funds provided for instruction have been diverted away from scholarships).

Accounting for a quadratic functional form in a regression model involves squaring the independent variable in question (whether in the data cleaning process or when running the regression model, similar to interaction terms)—in our case, instructional expenditure—and entering this squared value (X_i^2) into the regression model along with the original predictor (X_i), as illustrated in this equation:

$$Y_i = a + b_1 X_i + b_2 X_i^2 + e_i$$

If the coefficient on the main term (b_1) is negative and the coefficient on the squared term (b_2) is positive, this suggests a u-shaped relationship between the independent and dependent variables (the left-hand graph in Fig. 8.4). On the other hand, if the coefficient on the main term (b_1) is positive and the coefficient on the squared term (b_2) is negative, this suggests an inverse-u-shaped relationship between the independent and dependent variables (the right-hand graph in Fig. 8.4). As shown in the second column of Table 8.6, in our running example, the coefficient for instructional expenditure is significant and positive ($b = 12.91$, $p < 0.001$) and coefficient for the squared value of instructional expenditure is significant and negative ($b = -0.01$, $p < 0.05$), suggesting an inverse-u relationship between instructional expenditure and international student enrollment (we can also see this relationship visually in the graph on the right of Fig. 8.2).

We can also plot the residuals from this model that includes a quadratic functional form (the squared value of instructional expenditure). These residuals are shown in Fig. 8.5. This plot shows some improvement over the residuals from the linear model depicted in Fig. 8.3 in that the values are more densely and consistently clustered around zero (however, note that a couple of predictions are still off by around 4000 international students). Residuals for institutions predicted to enroll high numbers of international students also appear to be more evenly dispersed around zero. If I were including these regression models in a research paper,

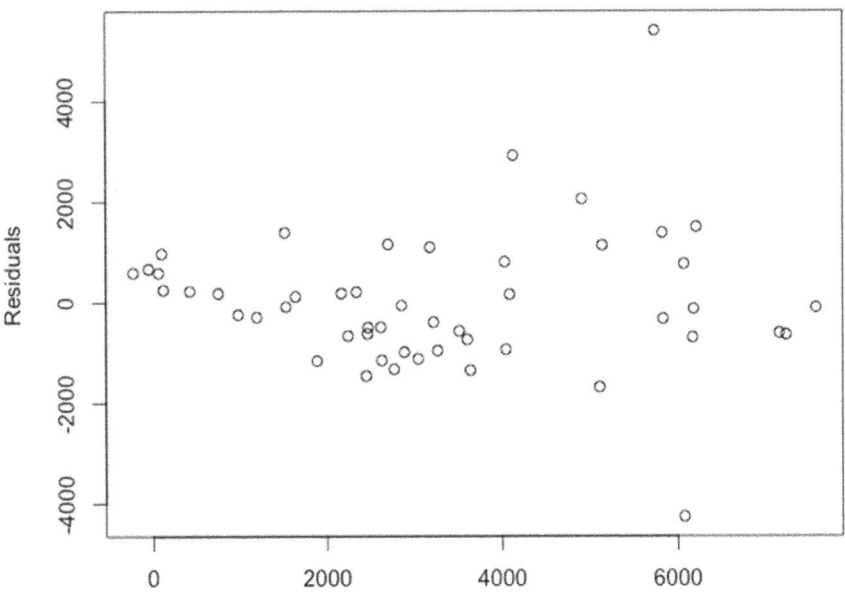

Fig. 8.5 Residuals from a quadratic functional form model using instructional expenditure to predict international student enrollment (*Data source:* Sample Dataset #1 - US National Center for Education Statistics, Integrated Postsecondary Education Data System)

I would probably share the results of both models with readers, but argue that the second one, which introduces the quadratic functional form for instructional expenditure, is a somewhat better fit for the data than the first.

8.4 Binary Outcome Variables

Thus far in our discussion of regression modeling, we have considered only examples where our outcome variable is continuous (e.g., international student enrollment). However, in international education research, we often care about outcomes that are binary (e.g., whether an institution offers study abroad opportunities). In this case, our dependent variable is a series of 0's and 1's, like the Hospital variable that we constructed earlier in this chapter. In our running example in this section, we will consider institutional characteristics that predict whether a US community college (defined as a public, two-year institution) offers study abroad opportunities to students. This variable is binary in that a 1 indicates that a college offers study abroad and a 0 indicates that it does not. The data that we will see in this section are found in Sample Dataset #3 and come from the US National Center for Education Statistics' Integrated Postsecondary Education Data System, representing the 2015–16 academic year. The institutions in this particular dataset represent all community colleges that reported data for the variables we make use of in this section in this academic year (N = 1,019).

Like the continuous outcome variables that we have been considering, a binary outcome variable, such as whether a community college offers study abroad, is numerical, even if it is just a column of 0's and 1's. We can use a binary outcome variable as the dependent variable in a linear regression model, just like we would for OLS. When an outcome variable is binary, linear regression produces what is called a **linear probability model** (LPM). To illustrate a LPM regression model, I ran the following simple linear regression:

$$Y_i = a + bX_i + e_i,$$

where the outcome (Y_i) is whether the community college offers study abroad (1 = yes, 0 = no) and the predictor (X_i) is total enrollment at the college (in 1000s). The results of this regression analysis are found in Table 8.7 and indicate that total enrollment has a positive and significant relationship with offering study abroad opportunities ($p < 0.001$). That is, community colleges with larger enrollments are more likely to offer study abroad. In an LPM such as this one, coefficients are interpreted as increases or decreases in the probability of the outcome variable. That is, the coefficient on total enrollment in this model indicates that for every 1000 increase in enrollment, the probability that a community college will offer study abroad goes up by 0.015.

LPMs are useful in that they are estimated in the exact same way as linear regression and as such are easy to implement and interpret for someone who is already familiar with OLS. However, the major drawback of LPM is that it allows

8.4 Binary Outcome Variables

Table 8.7 Linear probability model predicting whether a community college offers study abroad

Variable	
(Intercept)	0.008
	(0.011)
Total Enrollment (in 1000s)	0.015***
	(0.001)
N observations	1019
R^2	0.132
Adjusted R^2	0.131

Note Standard errors in parentheses
*** $p < 0.001$ (two-tailed p-values)
Data source: Sample Dataset #3 - US National Center for Education Statistics, Integrated Postsecondary Education Data System

for predicted values that are impossible. The left-hand plot in Fig. 8.6 illustrates this feature of LPM. Here, like before, the x-axis represents values of our predictor variable, total enrollment (in thousands), and the y-axis represents whether a community college offers study abroad. The circles represent data points. Logically, these data points cluster at 0 (the institution does not offer study abroad) and 1 (the institution does offer study abroad) along the y-axis. The line in this figure represents the predicted probability values from the LPM that we just estimated. Notice that at community colleges with very high enrollments, the predicted probability that an institution will offer study abroad is greater than 1—an impossible scenario (by definition, probabilities can only range from 0 to 1). We can see these impossible predicted probabilities more clearly in the residual plot of this

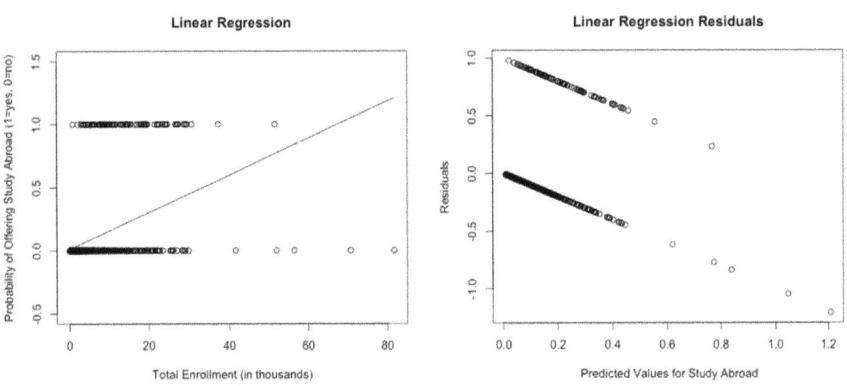

Fig. 8.6 Predicted probabilities resulting from a LMP model using total enrollment to estimate whether a community college offers study abroad (left) and the residuals from this model (right) (*Data source:* Sample Dataset #3 - US National Center for Education Statistics, Integrated Postsecondary Education Data System)

regression model on the right-hand side of Fig. 8.6, which shows two predicted probabilities that are higher than 1.

An additional property of LPM is that it assumes a constant linear relationship between the predictor variable (total enrollment) and the outcome variable (offering study abroad). This assumption suggests that when a small community college with only 1000 students doubles its enrollment (that is, it enrolls 1000 more students), its probability of offering study abroad is predicted to increase by the same amount that would be associated with a 1000 increase in enrollment at a community college with an enrollment of 10,000. This assumption is not logical—doubling enrollment from 1000 to 2000 may not be enough to support a study abroad program at a smaller community college, and once a community college reaches a certain size, it is likely to have enough enrollment to support study abroad, regardless of any future enrollment increases. Additional gains in enrollment would not do much to change the probability that a college with already large enrollments offers study abroad opportunities. To this end, we might expect enrollment increases in the middle rather than at the two extremes of the enrollment distribution (very low or very high enrollment) to have a more dramatic relationship with the probability that a community college will offer study abroad. These two characteristics of LPM—predicted probabilities that are impossible and the unrealistic expectation that the relationship between the predictor variable and the outcome variable is constant—suggest that a linear regression model is not the best option to estimate the relationship between our predictor variable and the binary outcome. In education research, the most common solution to these issues we have encountered is to use a different kind of regression modeling entirely—that is, **logistic regression**.

8.4.1 Logistic Regression

Logistic regression is a kind of regression modeling that is used specifically to model a dependent variable that has two possible values, like our study abroad offering outcome. In most statistical software programs, logistic regression is just as easy to implement as OLS and simply requires a small change in the command that produces the regression model. Table 8.8 summarizes the results of a logistic regression model that uses total enrollment to predict the likelihood that a community college will offer study abroad, just like the LPM that we have already estimated. Notice this model also predicts a positive relationship between total enrollment and whether a community college offers study abroad.

Instead of assuming a linear relationship between a predictor and an outcome variable, logistic regression assumes that the relationship between two variables is s-shaped, as illustrated in the plot in Fig. 8.7, which shows the predicted probabilities that a community college will offer study abroad, based on the logistic regression in Table 8.8. Notice how logistic regression restricts the probabilities of offering study abroad to fall between 0 and 1 so that predicted probabilities are

8.4 Binary Outcome Variables

Table 8.8 Logistic regression model predicting whether a community college offers study abroad

Variable	
(Intercept)	−3.195***
	(0.175)
Total Enrollment (in thousands)	0.118***
	(0.014)
N observations	1019
McFadden's Pseudo R^2	0.142

Note Standard errors in parentheses
*** $p<0.001$ (two-tailed p-values)
Data source: Sample Dataset #3 - US National Center for Education Statistics, Integrated Postsecondary Education Data System

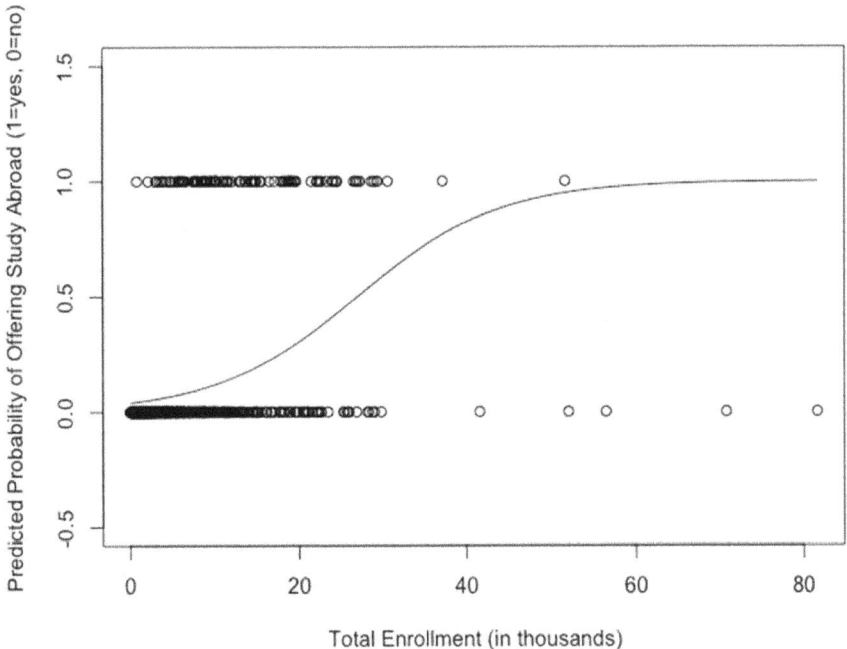

Fig. 8.7 Predicted probabilities resulting from a logistic regression model using total enrollment to estimate whether a community college offers study abroad (*Data source:* Sample Dataset #3 - US National Center for Education Statistics, Integrated Postsecondary Education Data System)

all realistic values. Logistic regression has the additional advantage of not assuming a linear relationship between a predictor and an outcome variable. The plot in Fig. 8.7 suggests that for colleges with low enrollment, the probability of offering study abroad is near 0 and that for colleges with high enrollment, the probability of offering study abroad is near 1. The real changes in probability of offering study abroad happen near the middle of the total enrollment distribution, where the slope of the regression line is quite steep.

So, how does logistic regression create this s-shaped curve that is much more realistic in terms of both predicted probabilities and the assumed relationship between the dependent and independent variables? The short answer to this question is that it transforms the outcome variable into something called **log odds**.[6] The mathematical operations behind log odds are beyond the scope of this book, but what this transformation does is give us the s-shaped curve in the predicted probability plot that we see in Fig. 8.7. A logistic regression equation looks much like an OLS regression equation, except that the outcome is written as the natural log (formally notated using 'ln') of the odds that a binary outcome will occur. In our example, this is the odds that a community college will offer study abroad. The logistic regression equation for our running example looks like this:

$$\ln\left(\frac{P_i}{1-P_i}\right) = a + bX_i.$$

Here, $\ln\left(\frac{P_i}{1-P_i}\right)$ is the natural log of the odds that a community college will offer study abroad. (As a related point of interest, P_i is the probability that a community college will offer study abroad. This, too, can be predicted by statistical software.) On the right-hand side of this equation, a is the intercept, just like in OLS, and X_i is our predictor variable—in this case, total enrollment. b is the regression coefficient associated with total enrollment—the predicted increase in the log odds of offering study abroad when enrollment increases by 1000. Notice that this equation does not have an error term. That is because, at its heart, logistic regression estimates something that is not observed—the probability that a particular scenario (in our case, offering study abroad) will occur. That is, we know whether a community college offers study abroad (or not), but we cannot know the real probability that a particular community college will provide this international opportunity to students. For this reason, we cannot compare predicted and observed values to calculate residuals like we would in OLS.[7]

While logistic regression has several properties that give it certain advantage over LPM when the outcome of interest is binary, interpretation of regression coefficients is one of the drawbacks of logistic regression. That is, we know that

---

[6] What this transformation does is force the relationship between the predictor variable and the outcome variable to be linear. In this way, even though the relationship between a predictor variable and the outcome variable is not linear, the relationship between a predictor variable and the *log odds* of the outcome variable is.

[7] In fact, logistic regression does not use a squared-residuals approach to estimate the regression line at all. Instead, this kind of regression modeling uses what is called maximum likelihood estimation. Maximum likelihood estimation takes an iterative approach to estimating regression parameters using a guided trial and error strategy. This approach essentially involves trying out different regression coefficients until the one that fits the observed data best—that is, the one where the observed data is most likely—has been discovered.

8.4 Binary Outcome Variables

a one-unit increase in total enrollment is 1000 students (because we measured this variable in 1000s), but what does the 0.118 regression coefficient on total enrollment in Table 8.8 mean? This is a 0.118 increase in the natural log of the odds that a community college will offer study abroad. This number has very little meaning in the world beyond statistics textbooks. To this end, we have to convert logistic regression coefficients so that we can interpret them in a way that makes sense in real-life situations. Two common ways of expressing logistic regression coefficients are **odds ratios** and **average marginal effects**. Like the log odds that result from a logistic regression themselves, these two conversions are mathematically complex. Luckily, computer software will make them for you. For this reason, here, we do not focus on how these conversions happen, but rather their interpretation.

8.4.2 Odds Ratio

While you had probably never heard of log odds before reading this book, you probably have heard of odds before, which is a common way to talk about the likelihood of an outcome, particularly in the world of gambling. Mathematically, the odds are the ratio of the probability that an event will happen divided to the probability that an event will not happen—$\frac{P_i}{1-P_i}$ in our logistic regression equation. If the probability that a community college will offer study abroad is 0.5 (P_i), the probability that it will not offer study abroad is 0.5 ($1 - P_i$), and the odds of the college offering study abroad are 1. (That is, 1-to-1 odds.) In this case, the college is just as likely to offer study abroad as it is not to. When we are interpreting odds, values above 1 mean that an event (in our case, the offering of study abroad) is more likely to happen, while values below 1 mean that an event is less likely to happen. If the probability that a community college will offer study abroad is quite high—say, 0.8 (P_i)—the probability that it will not offer study abroad is low, 0.2 ($1 - P_i$). In this case, the odds that the college will offer study abroad are very high: 4.0 (0.8/0.2). (That is, 4-to-1 odds.) On the other hand, if the probability that a community college will offer study abroad is low—say, 0.3 (P_i), then the probability that it will not is high, 0.7 ($1-P_i$), and the odds that it will offer study abroad are low, around 0.4 (0.3/0.7).

When a statistical software program converts the log odds that result from a logistic regression analysis into an *odds ratio*, it takes the exponential of the log odds regression coefficient.[8] Just like division is the opposite of multiplication, taking the exponential is the opposite of taking the log of a number—that is, the exponential *undoes* the logging of the odds so that we end up with an odds ratio that is more readily interpreted. It is important to note that while *odds* are themselves a ratio, an *odds ratio* is a ratio of ratios. Hence, this term tells us by

---

[8] A statistical software program will also take the exponent of the coefficient's standard error, thus accounting for the conversion from log odds to odds for the purposes of significance testing.

what factor the odds change. When I directed my statistical software program to take the exponential of 0.118, the expected increase in the log odds of offering study abroad associated with an increase of 1000 enrolled students, I got an odds ratio of 1.125. This means that for a 1000 student increase we would expect a 12.5% increase—referring to the numbers after the decimal point in 1.125—in the odds of a community college offering study abroad. When the odds ratio is lower than 1, instead of referring to the numbers after the decimal point (subtracting 1), we subtract the odds ratio from 1 for the purpose of interpretation. For example, if an odds ratio is 0.125, we subtract this value from 1 to arrive at 0.875. This value is then interpreted as an 88% decreased likelihood of whatever the outcome variable is.

8.4.3 Average Marginal Effects

Average marginal effects are another option for converting log odds so that they are easier to interpret, and many researchers find them easier to work with than odds ratios because of their simple interpretation. Like the odds ratio, average marginal effects are relatively easy to calculate in most statistical software programs.[9] As their name suggests, the average marginal effect is calculated by averaging the discrete changes in the predicted probability of a given outcome (in our case, offering study abroad) when independent variables are set at all possible values for all observations in a dataset. The average marginal effect for a given variable indicates the average change in the probability of the dependent variable (in our case, the change in the probability that a community college will offer study abroad) associated with a one-unit change in a given dependent variable (in our case, an increase in 1000 students enrolled). When I calculated the average marginal effect for total enrollment corresponding to the logistic regression model in Table 8.7, I ended up with a very small number: 0.00000854. This number indicates that with an increase of 1000 students, the probability that a community college would offer study abroad increases by about 0.000854 percentage points (notice that the decimal point moves to the right two places in this interpretation) or by 0.00000854 on a proportion scale. This is not a very large increase in the probability of offering study abroad.

8.4.4 Pseudo-R^2

A final piece that is missing from our discussion of logistic regression is a measure of model fit. Recall in Chap. 7 that we introduced R^2 as a measure of model fit

---

[9] However, calculations of marginal effects are not currently available in the most recent version of a standard SPSS installation. The software programs that will calculate marginal effects (such as Stata or R) will also convert the standard error for the purpose of significance testing of regression coefficients.

for OLS regression, describing how this measure represents the percentage of the variation in the dependent variable that the independent variables explain. Because logistic regression is calculated differently from OLS regression, a similar measure of model fit is not available. Instead, researchers can choose among several measures that are similar to an OLS R^2 that apply to a logistic regression scenario, called pseudo-R^2 calculations. McFadden's pseudo-R^2 is a common measure found in the literature, but whichever choice you use be sure to mention the name of the measure when you report results.[10] In general terms, McFadden's pseudo-R^2 is calculated by comparing the likelihood of the logistic regression model that a researcher estimates to the likelihood of a null model (a logistic regression model without any predictors). In this case, the word *likelihood* refers to the likelihood of obtaining a given regression model (i.e., specific regression coefficients) in the context of a specific dataset. A high McFadden's pseudo-R^2 indicates a better model fit. Like R^2 from an OLS regession model, McFadden's pseudo-R^2 ranges in value from 0 to 1. In our running example, McFadden's pseudo-R^2 was 0.142 (see the last row of Table 8.8). Given that our logistic regression contained only one predictor variable, this lower value is not especially surprising.

8.5 Example from the Literature

We end this chapter with an example of a study that incorporates dummy variables as predictors of interest in a logistic regression model (Lingo 2019), thus combining several of the concepts from this chapter. In this study, the author uses data from the 2006–2012 years of the Wabash National Study on Liberal Arts Education (a U.S.-based dataset) to explore the relationship between a student's parents' educational attainment level and whether the student participates in study abroad. (Note that this study uses the U.S. definition of *study abroad*, meaning that a student who is enrolled at a U.S. institution of higher education spends a part of their studies in a foreign country—whether enrolled at a foreign institution or enrolled in a home-institution sponsored program that takes place abroad.) Theoretically, this study takes a social stratification approach to explain disparities in study abroad participation, hypothesizing that students whose parents have higher levels of educational attainment will be more likely to study abroad. Analytically, Lingo (2019) uses logistic regression to predict participation in study abroad (a binary variable, 1 = yes, 0 = no) by the end of a student's fourth year of higher education enrollment and uses dummy variables to indicate parental educational attainment. This study's most robust logistic regression model takes the form of the following equation:

$$\ln\left(\frac{P_i}{1-P_i}\right) = a + X_{1i}b_1 + X_{2i}b_2$$

---

[10] Other common pseudo-R^2 measures include Efron's pseudo-R^2 and Nagelkerke's pseudo-R^2.

Table 8.9 Logistic regression model predicting study abroad participation, intercept and control variables omitted (Lingo, 2019, p. 1158 and p. 1160)

Predictor	Odds ratio	Predicted probability of study abroad
Less than bachelor's degree	Comparison Group	0.389 (0.022)
Bachelor's degree	1.125 (0.126)	0.411 (0.021)
Advanced degree	1.531 (0.141)***	0.467 (0.021)***
N	3824	3824
Pseudo-[11]R^2	0.221	0.221

Note Standard errors in parentheses
*** p < 0.001 (two-tailed *p*-values)

In this equation, the predictors of interest are represented in X_{1i}, which depicts three levels of parental education (less than a bachelor's degree, bachelor's degree, and advanced degree). The additional variables in this model represent control variables (X_{2i}): a student's intent to study abroad in their first year of enrollment, whether the student identifies as female, the student's racial/ethnic identity, whether the student received a grant, whether the student received a loan, the student's standardized test score at enrollment, the kind of institution a student attended (regional university, research university, or liberal arts college), the student's field of study, whether the student intended to earn an advanced degree, a student's first-year academic achievement (grade point average [GPA]), the number of diversity-focused courses a student took, a measure of the student's openness to diversity, a measure of the student's non-classroom interaction with faculty, hours a student spent on co-curricular activities, hours a student spent working, hours a student spent studying, and hours a student spent socializing. Notice that in the equation above, I have represented this very long list of control variables as a single vector (X_{2i}) to focus attention on the predictor of interest (X_{1i}) and to make the equation easier to read.

Table 8.9 summarizes Lingo's results corresponding to this equation. Here, I provide only the results for the predictor of interest (parental education) and leave out results corresponding to control variables (but it is important to note that these variables are in the equation). Notice that these predictor variables are dummy variables, and that when odds ratios are presented (the middle column), the category 'Less than bachelor's degree' is used as the comparison group or reference category. These results suggest that the odds that a student whose parents had obtained an advanced degree would study abroad were 53% greater than a student whose parents had obtained less than a bachelor's degree ($p < 0.001$). Similarly, when we look at the column of predicted probabilities, we see that the probability that a student whose parents had obtained an advanced degree would study

---

[11] Note that Lingo (2019) does not provide information about hte kind of Pseudo-R reported in this study.

abroad was around 0.08 (the average marginal effect) greater compared to a student whose parents had obtained less than a bachelor's degree (0.467–0.389 = 0.078; $p < 0.001$). Another way of expressing this result is that a student with advanced-degree parents was around eight percentage points more likely to study abroad compared to a student whose parents had completed less than a bachelor's degree.

8.6 Looking Forward

The regression techniques presented in this chapter and Chap. 7 are powerful and allow researchers to explore many different types of relationships among variables in a particular dataset. This chapter has focused on different kinds of variables that can be used as predictors in regression models (dummy variables and interaction effects), different shapes that can depict the relationships among variables (functional form), and ways to fruitfully explore outcome variables that are binary (LPM and logistic regression). However, we end this chapter with the same caveat as the previous one—these relationships are still correlational rather than causal. In education research, we are often concerned with whether one variable causes another—such is whether a newly-designed study abroad program causes students to graduate faster or whether a new scholarship program entices more students to participate in international education. The next chapter draws on concepts introduced in this chapter and Chap. 7 to describe several approaches to quantitative analysis that approach a causal interpretation.

8.7 Practice Problems

1. Sample Dataset #2 contains information about 161 Liberal Arts institutions in the 2016–17 academic year. In this first practice activity, we focus again on the number of students who studied abroad from these institutions (studyabroad). The table below summarizes an OLS regression model that builds on the one introduced in practice activity 1 in the previous chapter. This regression model adds three variables to this equation: one that allows the acceptance rate variable to depict a quadratic functional form ($AcceptRate_i^2$), a dummy variable indicating whether the institution is public ($Public_i$) and a final variable that represents the interaction between $Public_i$ and $TotalEnroll\ 1000_i$ ($Public * TotalEnroll\ 1000_i$).

$$TotalSA_i = a + b_1 TotalEnroll\ 1000_i + b_2 PctFemale_i + b_3 AcceptRate_i \\ + b_4 AcceptRate_i^2 + b_5 Public_i + b_5 Public * TotalEnroll\ 1000_i + e_i$$

Variable	Coefficient
(Intercept)	137.85*
	(55.89)
Total Enrollment (in 1000s)	116.77***
	(11.55)
Percent female	−0.93+
	(0.54)
Acceptance rate	−3.21+
	(1.70)
Acceptance rate squared	0.02
	(0.02)
Public	87.95+
	(50.09)
Public*Total Enrollment (in 1000s)	−98.83***
	(14.79)
N observations	161
R^2	0.48
Adjusted R^2	0.46

Note Standard errors in parentheses
+$p < 0.10$, *$p < 0.05$, ***$p < 0.001$
Data source: Sample Dataset #2 - US National Center for Education Statistics, Integrated Postsecondary Education Data System

a. Is there evidence that the functional form depicting the relationship between acceptance rate and study abroad participation is quadratic? Why or why not?
   i. What diagnostic test might you conduct to determine whether this functional form is appropriate for your data?
b. How do you interpret the coefficient corresponding to $Public_i$?
c. How do you interpret the coefficient corresponding to $Public * TotalEnroll\ 1000_i$?
   i. What might you do to explore further the nature of this interaction term?
d. What percentage of the variance in study abroad participation does this regression model explain?

2. Now use your statistical software program of choice to load Sample Dataset #2 and run this regression model yourself. Double-check your work to ensure that you arrived at the same numbers as in the table above. In a second step, consider a quadratic functional form to depict the relationship between total enrollment and study abroad participation using a simpler version of this regression model. Estimate this model using the following equation:

$$TotalSA_i = a + b_1 TotalEnroll\ 1000_i + b_2 TotalEnroll\ 1000_i^2$$

8.7 Practice Problems

$$+ b_3 PctFemale_i + b_4 AcceptRate_i + b_5 Public_i + e_i$$

a. Interpret the coefficient corresponding to total enrollment and total enrollment squared. Is there evidence that the relationship between total enrollment and study abroad participation is quadratic? How do you know?
   i. What diagnostic test might you conduct to determine whether this functional form is appropriate for your data?
b. What lessons might international educators derive from the regression models in practice problems 1 and 2?

3. Sample Dataset #3 contains the community college data used to derive the logistic regression examples used in this chapter. Use this dataset to build on the regression model in Table 8.7, which takes offering study abroad as the outcome of interest (studyabroad_offered), so that it accounts for three additional predictor variables (Notice that this regression model will contain 915 observations due to missing data on tuition charges and locale from some community colleges. Be wary of missing data when you create variables for this exercise). Namely, estimate a logistic regression model that takes the following form:

$$\ln\left(\frac{P_i}{1 - P_i}\right) = a + b_1 TotalEnroll\_1000_i + b_2 Rural_i$$

$$+ b_3 PctOver25_i + b_4 Tuition\_1000_i$$

In this equation, $Rural_i$ is a binary indicator of whether the institution is located in a rural area (this variable can be derived from locale—see the Data Dictionaries at the beginning of this book for variable definitions), $PctOver25_i$ is the percentage of the student population aged 25 or older (pctover25), and $Tuition1000_i$ is the amount charged in tuition at each college in US$1000 units (this variable can be derived from tuition).

a. Convert the log odds coefficients from this regression model into odds ratios. How do you interpret the odds ratios corresponding to total enrollment, rural, percent 25 + and tuition charges?
b. Convert the log odds coefficients from this regression model into average marginal effects. How do you interpret the probability changes corresponding to total enrollment, rural, percent 25+ and tuition charges?
c. How well does this regression model fit the data? How do you know?
d. What might international educators at community colleges learn from this regression model?

Recommended Reading

A Deeper Dive

Gujarati, D. N., & Porter, D. C. (2010). Functional forms of regression models. In *Essentials of Econometrics* (pp. 132–177). McGraw-Hill.

Gujarati, D. N., & Porter, D. C. (2010). Dummy variable regression models. In *Essentials of Econometrics* (pp. 178–216). McGraw-Hill.

Wooldridge, J.M. (2016). Multiple regression analysis: Further issues. *Introductory Econometrics: A Modern Approach* (pp. 166–204). Cengage Learning.

Wooldridge, J. M. (2016). Multiple regression analysis with qualitative information: Binary (or dummy) variables. *Introductory Econometrics: A Modern Approach* (pp. 205–242). Cengage Learning.

Additional Examples

Di Pietro, G. (2020). Does an international academic environment promote study abroad? *Journal of Studies in International Education*, 1028315320913260.

Finn, M., Mihut, G., & Darmody, M. (2021). Academic Satisfaction of International Students at Irish Higher Education Institutions: The Role of Region of Origin and Cultural Distance in the Context of Marketization. *Journal of Studies in International Education*, 10283153211027009.

Haupt, J. P. (2021). Short-Term Internationally Mobile Academics and Their Research Collaborations Upon Return: Insights From the Fulbright US Scholar Program. *Journal of Studies in International Education*, 1028315321990760.

Martinsen, R. A. (2010). Short-term study abroad: Predicting changes in oral skills. *Foreign Language Annals*, *43*(3), 504–530.

9

Introduction to Experimental and Quasi-Experimental Design

In several chapters of this book, words of caution have been given about how correlation is not the same as causation. Even the most well-designed regression model is unable to provide evidence that a particular cause produces a specific effect if a study is not designed for causal inference. The studies and examples provided in this book thus far can be described as **observational studies**, meaning that they were conducted with the purpose of *describing* a particular international education-related phenomenon. For example, the Lingo (2019) study used as an example in Chap. 8 aimed to describe the relationship between parental education (the predictor variable) and study abroad participation (the outcome variable). It would be wrong to conclude from this study that having highly educated parents causes a student to study abroad. Indeed, many other explanations for the association between parental education and study abroad participation are possible. For example, educational attainment is often related to earning a higher salary. Students whose parents have higher levels of educational attainment likely come from homes that are more economically prosperous, thus facilitating study abroad participation. Another possible explanation is that highly educated parents are more likely to have had abroad experiences themselves, and therefore encourage their children to study abroad to a greater extent than parents who have not traveled abroad. In either case, the relationship between parental educational attainment and study abroad participation is one of association rather than causation. Research that is designed for **causal inference**, on the other hand, allows us to isolate, at least in theory, the causal link between a treatment condition and a specific outcome.

Causal inference refers to a family of analyses, and this group of analytic approaches is the topic of this chapter. This chapter is different from previous chapters in that it is intended to introduce readers to several common causal inference research designs that are useful in international education research. The main idea is that by the time you finish this chapter, you will understand the basic logic and role of each of these research designs. The purpose of this chapter is not to provide you with all the tools you need to conduct one of these analyses. If you decide you want to carry out a study using one of the research designs outlined in this book, further reading recommendations are found at the end of the chapter. Specific research designs that this chapter covers include randomized control trials, propensity score modeling, regression discontinuity, and difference-in-differences.

9.1 Experimental Design: Randomized Control Trials

Randomized control trials (RCTs) represent perhaps the most conceptually straight-forward research design for causal inference. It will soon become clear that they are also often the most difficult to carry out in practice, and for many research questions RCTs are not even appropriate. These studies take advantage of the properties of samples produced through randomization. Recall from Chap. 1 that random sampling involves selecting units from a population so that each unit has equal likelihood of being selected into the sample. Random sampling is important because, in the absence of sampling error, data collected in this way are representative of the population and are not biased in any way in favor of one group or another. In the context of an RCT, study participants are randomly assigned into two groups: a treatment group and a control group. Because these groups are randomly assigned, in the absence of sampling error, they are considered equivalent to one another.

9.1.1 Treatment and Control Groups

Treatment and control groups are perhaps best defined based on what happens to units in each of the groups *after* randomization occurs. In brief, the treatment group receives some sort of intervention, while the control group continues in what is often referred to as the *business-as-usual* condition. Interventions can be loosely defined as an educational context or activity that the treatment group experiences, and the control group does not. In international education, and in education broadly, interventions can be found everywhere. A policy that provides study abroad scholarships to one group of students and not another is an intervention—students who received scholarships comprise the treatment group and those who did not comprise the control group. Introduction of a particular internationalized curricular approach is another example of an international educational intervention. In this case, students who are exposed to the curricular approach are in the treatment group and those who are not exposed are in the control group.[1] The running example that we will use in this chapter involves the introduction of an intercultural learning workshop on a college campus. Students who participate in this workshop are the treatment group while students who do not participate

---

[1] The extent to which randomization is possible in these scenarios is an important consideration that we will discuss later in this chapter.

9.1 Experimental Design: Randomized Control Trials

are in the control group. For the purpose of illustration, we will assume that for students randomly assigned to participate in the intercultural learning workshop are required to attend—that is, they cannot refuse to participate. In reality, situations like this one are rarely the case in international education research (anyone who has worked with students knows that it is hard to coerce students to do things they do not want to do, never mind the ethical implications of such a situation), representing one of the drawbacks of randomized control trials in education research.

9.1.2 Randomization and Selection Bias

An important caveat to our discussion about treatment and control groups thus far is that randomization of units into these two groups has to happen *before* the intervention. When we randomly assign students to treatment and control groups, we ensure that, as long as randomization works like it should, both groups will be similar in terms of both **observed** and **unobserved variables** so that the only difference between the groups is the intervention itself. Both observed and unobserved variables, or characteristics, have the potential to influence whether a participant would choose to be in the treatment or control group if left to their own devices. By observed characteristics, I mean information about study participants that we either know or could find out if we needed to. In our running example, student demographics are an excellent illustration of observed characteristics. This information is typically collected on college campuses. Unobserved characteristics, on the other hand, are characteristics that are difficult or even impossible to observe. An unobserved characteristic that may be especially relevant in our running example is whether a student enjoyed their previous intercultural interactions. While we might be able to administer a survey to collect this kind of information, measuring something like "enjoyment" is difficult, and there is no guarantee that students will be able to recall all their previous intercultural interactions in the moment that they are taking a survey. In this case, random assignment to treatment and control groups is key to the success of the study. Because students have an equal likelihood of belonging to the treatment or control groups, randomization ensures that the groups are balanced on both observed and unobserved characteristics. Table 9.1 provides an illustration of this concept.

In Table 9.1, we see that treatment and control groups are comparable regarding the mean number of hours students study per week, mean GPA, the percentage of students in each group that had previously studied abroad, and the percentage of students eligible to receive Pell funding, an imperfect but commonly used indicator of a student's socioeconomic status. These characteristics are all observable. If we were unconvinced that the two groups were comparable for a given characteristic (e.g., the two groups do vary some in terms of mean hours studied per week), we could run the appropriate statistical test (e.g., a *t*-test) to determine if the difference between group means is significantly different from zero. Table 9.1 also shows that these two student groups are similar in terms of how much they enjoyed

Table 9.1 Intercultural workshop participant information—randomization

Characteristic	Treatment group	Control group
Mean hours studied per week	16.01	15.84
Mean GPA	3.21	3.25
Prior abroad experience (percent yes)	31%	29%
Pell eligibility (percent yes)	43%	45%
Intercultural interaction enjoyment (Scale: 1–4)	*3.6*	*3.5*

Note Italics indicate that intercultural interaction enjoyment is unobserved. Data in this table are invented for illustrative purposes

their previous intercultural interactions, a characteristic that is, in this example, unobserved (indicated in the table using italics). Randomization of students into the two groups ensured that this would be the case. In experimental contexts, we make the assumption that treatment and control groups are balanced regarding both observed *and* unobserved characteristics. We can—and should—check to ensure that randomization produced balanced groups for observed characteristics, but we have to make this assumption regarding unobserved characteristics.

In ensuring that treatment and control groups are balanced on observed and unobserved characteristics, randomization avoids introducing **selection bias** into a study. Table 9.2 provides the same information given in Table 9.1 but illustrates what might happen if students were able to choose, or select, whether they participated in the intercultural workshop. As this table shows, students in the treatment group studied more on average per week, had a higher mean GPA, were more likely to have studied abroad, and were somewhat less likely to be eligible for a Pell grant. Additionally, regarding our unobserved characteristic, treatment group students on average enjoyed their previous intercultural interactions more than control group students. This previous enjoyment of intercultural interactions might be responsible for treatment group students' interest in the intercultural workshop in the first place.

Table 9.2 Intercultural workshop participant information—naturally occurring groups

Characteristic	Treatment group	Control group
Mean hours studied per week	20.34	7.45
Mean GPA	3.56	3.01
Prior abroad experience (percent yes)	63%	18%
Pell eligibility (percent yes)	42%	47%
Intercultural interaction enjoyment (Scale: 1–4)	*3.8*	*2.1*

Note Italics indicate that intercultural interaction enjoyment is unobserved. Data in this table are invented for illustrative purposes

9.1.3 Outcomes

Thus far in our running RCT example we have introduced the intervention, or the cause of change, for students in the treatment group—the intercultural workshop—but we have yet to define the effect—the outcome. The importance of balance between treatment and control groups becomes clear when we consider the change that we would expect to produce with an intervention. Growth in intercultural competence is a clear effect that we might expect to observe among students who participate in an intercultural learning workshop. For the sake of illustration, let us suppose that at the end of this workshop, we administered a measure of intercultural competency to both the treatment and the control groups. On a scale of 1–5, students in the treatment group scored 4.85 and students in the control group scored 3.46. We conduct a t-test of these differences in means and discover that this difference is statistically significant at $\alpha = 0.05$.

In the case where students were randomized into treatment and control groups, illustrated in Table 9.1, we could reasonably conclude that the intercultural learning workshop was the cause of students' advancement in intercultural competency—randomization has ensured that the only meaningful difference, both observed and unobserved, between the treatment and control groups is participation in the workshop. However, in the situation illustrated in Table 9.2, where students were able to select into workshop participation, multiple explanations for this result are possible. For example, students in the treatment group spent more hours studying on average before participating in the intercultural workshop, so maybe this student group also spent more time interacting with workshop materials outside the classroom. Another explanation is that treatment students were more likely to have previously studied abroad, so perhaps they began the workshop with higher levels of intercultural competency. In either case, it is not the workshop that produced the significant difference between treatment and control groups, but rather characteristics of the students themselves that were different for the two groups even before the workshop took place.

While we could account for these two potential explanations in a regression context, entering both hours spent studying per week and study abroad participation as control variables, thus partialling out their relationship with intercultural competence, we cannot do the same for unobserved characteristics, such as how much a student enjoyed previous intercultural interactions. This unobserved characteristic represents an additional explanation for a significant difference between the treatment and control groups on the post-intervention measure of intercultural competency. That is, students in the treatment group may have scored higher on the intercultural competency measure not because of participation in the workshop, but because they enjoyed intercultural interactions more than the control group. The extent to which these unobserved characteristics rather than the intervention itself explain a study's outcome variable shows up as bias in our results when participants are able to choose to participate in an intervention. As you might imagine, in some cases, selection bias is minimal, but in other cases, this bias can be quite large. If selection bias is large, our results are inaccurate, and are therefore of limited use for decision-making and generalizations to broader international education issues.

9.1.4 Threats to Validity

While the RCT is often considered the gold standard in education research because of its ability to eliminate selection bias and identify causal relationships in a relatively straightforward way, the integrity of an RCT hinges on several details going exactly right in the implementation of the study's design. While there are numerous threats to validity that can occur in an RCT, here I review three that are fairly common: **contamination**, **attrition**, and lack of **fidelity of implementation** of the intervention.

Contamination occurs when individuals belonging to treatment and control groups are able to interact with one another and share information amongst themselves. For example, consider what might happen if one of the treatment students in our running example was very close friends with one of the control students. This treatment student might share materials or discuss content from the intercultural workshop with their friend from the control group. If these friends spend a considerable amount of time together, this interaction has the potential to result in intercultural competence growth for the control student, even though this student was not exposed to the treatment condition. In this situation, contamination can lead to a diluted treatment effect. That is, our RCT would detect a smaller or even non-existent impact of the intercultural workshop on students' intercultural competence even though such an impact may very well exist.

An additional threat to the validity of an RCT is attrition. Attrition happens when students who were originally supposed to participate in a study drop out in a way that is not random. For example, students in the treatment group in our running example who are not especially interested in international topics or themes might stop attending the intercultural workshop, leaving this group to be comprised primarily of students who were interested in international topics to begin with. The attrition of these students from the study would leave treatment and control groups unbalanced along certain characteristics, many of which may be unobservable, thus leaving us in a situation more like Table 9.2 rather than Table 9.1. In this case, we could not count on the balance produced by randomization at the beginning of our study to be valid at the end when we measure our outcome of interest.

Finally, fidelity of implementation of the intervention is important for the results of an RCT to be valid. Fidelity of implementation refers to the extent to which a given intervention is implemented as intended. Consider, for example, a situation wherein a group of facilitators were responsible for implementing our intercultural workshop. For the sake of illustration, let us say that students in the treatment group were placed in one of five groups, each with a different facilitator. Some variation in how each facilitator implements the workshop would be expected. However, a situation where two or three facilitators decide to significantly modify workshop content or even not implement the workshop content at all could arise.

In this situation many if not most of the treatment group students are not receiving the treatment intervention as intended. To this end, researchers should have a plan in place to monitor and even measure the extent to which an intervention is implemented among treatment group participants.

9.1.5 Example from the Literature

While RCTs are not an especially common research design in international higher education, given the complications just outlined, they are conceptually important for several more complex research designs that are particularly useful in the field (reviewed in the next sections—Nonexperimental Contexts and Quasi-Experimental Design). An example of a study in international higher education that employs an RCT design is Meegan and Kashima's (2010) study exploring the emotional and self-esteem consequences of perceiving discrimination among international students. This study examines this topic using data from Asian international students enrolled at an Australian postsecondary institution. While Meegan and Kashima (2010) consider the impact of two different treatment conditions in their larger study, the discussion here focuses on their manipulation of a student's perception of the pervasiveness of discrimination against international students. In this part of the study, students were randomly assigned to treatment and control groups and were then asked to read an essay about discrimination against international students. Students in the treatment group read an essay that led them to believe that discrimination against international students was frequent and pervasive, while those in the control group read an essay about how discrimination against international students was rare. To measure their outcomes of interest, Meegan and Kashima (2010) asked participants to complete a five-item assessment of depressed affect and a ten-item self-esteem inventory. Importantly, the tasks that participants completed appeared in the order just described: Students were first randomized into treatment and control groups; then, manipulation of the experimental context took place (i.e., students were provided with a specific reading passage, based on their experimental group assignment); and finally, outcomes were measured.

To analyze their data, Meegan and Kashima (2010) used ANOVA analyses, such as those discussed in Chap. 5.[2] ANOVA was an appropriate methodological choice for this study because randomization of students into treatment and control groups ensured that the only difference between these two groups was their treatment or control group status. That is, there was no need to control for additional variables, such as students' demographics or academic information, as we have been doing in regression analyses, because these characteristics were evenly distributed across

---

[2] Note that here I report only a subset of Meegan and Kashima's (2010) results as I focus on only one of their two experimental conditions.

the treatment and control group contexts. The results of this study indicated no significant difference between the treatment and control groups regarding depressed affect or self-esteem. These results were in contrast to the authors' expectations of a negative impact of perceived discrimination on both outcomes. Should we assume that international students are immune to the effects of discrimination? This is likely not the case, but these results do raise important questions about exactly how experiencing discrimination affects international students. In fact, in this study, additional factors, especially the extent to which international student status was central to a student's identity, were shown to mediate the relationship between experimental conditions and the two outcomes of interest (depressed affect and self-esteem). In other words, an international student's identity appears to be especially salient in how they are impacted in contexts of discrimination.

9.2 Nonexperimental Contexts

While RCTs such as the one that Meegan and Kashima (2010) conducted may be the most straightforward approach to causal inference, these studies are not without their difficulties. In many situations in education research, randomization is simply not possible. Randomization is costly and requires a substantial time investment. Other times, specific rules or regulations mean that randomization is not allowed or is even illegal. Even when randomization is possible, it is often unethical. For example, if we truly believe that studying abroad is beneficial for students, it would be unethical to knowingly withhold the possibility of participation from a group of students. An equal ethical concern would be whether it is appropriate to force a student to study abroad. For this reason, we often find ourselves working with data that are collected through convenience samples or working with **secondary data** (also referred to as **observational data** or **field data**)—that is, data that were collected for other purposes, but that can be used to inform our research questions. Examples of secondary data include institutional datasets collected by offices of institutional research or institutional effectiveness, data collected by non-profit or governmental organizations, and data collected by entities such as UNESCO or the World Bank.

When we ask causal questions but intend to use nonexperimental data to answer them, selection bias becomes a serious issue that we have to contend with. While controlling for other variables that could impact the outcome of interest in a regression model is one way to minimize bias, it is highly unlikely that any dataset contains all the information that we would need for truly unbiased estimates of a treatment effect. That is, there will always be unobserved variables lurking in the background. Our previous example of how much a student enjoyed previous intercultural interactions is an excellent example of an unobserved variable that is likely not available in most secondary datasets. Other examples of variables that are unobserved, and that would be difficult to measure even if they were observed, include a student's motivation to achieve academically, an institution's

9.2 Nonexperimental Contexts

culture for international education, or the likelihood that a student will experience a racist incident in a specific host country. These unobserved characteristics have the potential to impact both whether a student participates in an international education intervention *and* the outcome of that intervention.

In this section, I introduce one way in which researchers often attempt to recreate experimental contexts using observational data, propensity score matching. However, I note from the beginning that, for a causal interpretation of results to be appropriate, this approach relies on the assumption that observed variables in the dataset are all that are needed to explain whether an individual chooses to participate in an intervention. In more technical terms, this assumption is that there are no unmeasured confounders that should be included in a study's analysis. In most cases, this assumption is tenuous at best.

9.2.1 Propensity Score Matching

Matching methods are a family of analytic approaches rather than a single approach to data analysis. Broadly speaking, **propensity score matching** relies on matching treatment and control students with one another in a way that reduces the differences between the two groups along observed characteristics. In this section, I focus on a rather standard propensity score matching approach (Rosenbaum & Rubin, 1985). However, it is important to be aware that matching techniques other than the one I present here are common and have the potential to produce estimates with less bias than the one I outline.[3] Propensity scores like the ones described in this section can also be used to derive weights that reduce bias in regression estimates of a treatment effect. A discussion of the various ways to approach matching or propensity score analysis is well beyond the scope of this book, but I have included references in the recommended reading section at the end of this chapter for readers who are interested in learning more about these techniques.

The goal of propensity score matching is to match similar treatment and control units so that the two groups are balanced on all observed characteristics—essentially recreating what would have happened if units had been randomized into treatment and control groups. To illustrate this concept, we explore the example dataset displayed in Table 9.3. We have two pieces of information about the twelve students that appear in this dataset: their status as treatment or control units and their score on an intercultural competence instrument that ranges from 1 to 4. A *t*-test of the difference in means between the treatment and control groups indicates that the groups are unbalanced in terms of intercultural competence, meaning that there is a significant difference between them ($t = -2.82$, $p < 0.05$). The mean

---

[3] Advanced readers who are interested in these alternatives can read more about exact matching (Rubin, 1973), genetic matching (Diamond & Sekhon, 2013), or coarsened exact matching (Iacus, King, & Porro, 2011).

Table 9.3 Example data for propensity score modeling

	Treatment (T) or Control (C)	Intercultural competence
Student 1	T	3.5
Student 2	T	3.9
Student 3	C	1.4
Student 4	C	2.6
Student 5	C	3.4
Student 6	T	4.0
Student 7	T	2.9
Student 8	T	1.7
Student 9	C	1.0
Student 10	C	1.2
Student 11	T	4.0
Student 12	C	1.3

Note Data in this table are invented for illustrative purposes

intercultural competence measure for the treatment group is 3.33 while the mean for the control group is 1.82.

If the purpose of our study is to explore the impact of studying abroad (our treatment) on students' gains in intercultural competence, and the measure of intercultural competence we have in our dataset is a pre-test score, meaning that it represents students' scores before studying abroad, we have a problem. That is, students in the treatment group already have significantly higher intercultural competence scores compared to those in the control group before they study abroad. How would we know that any post-treatment differences between the two groups are due to the treatment condition and not pre-treatment differences between the two groups?

The propensity score solution to this problem is to match students in the treatment group to similar students in the control group (or vice versa) to create a sample that is balanced on pre-treatment characteristics. In this case, we could match students in the treatment group to control group students with similar intercultural competence scores, as shown in Table 9.4. In this table, a student's intercultural competence pre-score is in parentheses to facilitate comparison

Table 9.4 Matched treatment and control groups (intercultural competence score in parentheses)

Treatment	Control
Student 1 (3.5)	Student 5 (3.4)
Student 8 (1.7)	Student 3 (1.4)
Student 7 (2.9)	Student 4 (2.6)

Note Data in this table are invented for illustrative purposes

9.2 Nonexperimental Contexts

between the two groups. Notice that here, we have matched Student 1, whose intercultural competence pre-score was 3.5, with Student 5, whose pre-score was 3.4. These are not exact matches between treatment and control students, but they are close. When I ran another t-test comparing treatment and control students in this matched sample, I found that the two groups were not significantly different from one another ($t = -0.30$, $p > 0.05$). The mean intercultural competence score for the treatment group was 2.7 while the mean for the control group was 2.5.

You are likely asking yourself what happened to the other six students in our dataset. The short answer to this question is that we simply removed them from our sample because we were unable to match them. The more technical answer is that they fell outside the area of **common support** between the treatment and control groups.[4] The area of common support refers to the area where treatment and control group participants' scores (in this case, their pre-test intercultural competence scores) overlap—that is, the area where matching between units can happen. Put differently, without common support, we cannot move forward with propensity score matching because there are no treatment group participants that can be appropriately matched to control group participants and vice versa. Researchers often explore the area of common support visually as a step to ensure that propensity score modeling is possible with a specific dataset. The area of common support for our dataset of pre-test intercultural competence scores is pictured in Fig. 9.1. Where the distributions of the treatment (the solid line) and control (the dashed line) groups overlap is where we were able to match students.

While our running example thus far has considered treatment and control groups that differ along the lines of a single variable, this example is not especially realistic. When treatment and control groups arise naturally from data rather than through randomization, they often differ from one another on multiple pretreatment variables. In this case, we would want to match treatment and control group students using all the variables that we think are important for determining whether a participant chose to belong to the treatment or control group. Matching treatment and control units on multiple variables can get very complex and becomes an impossible task very quickly (a somewhat dramatic term for this situation that is used in the propensity score literature is the *curse of dimensionality*, meaning that data with many variables are multi-dimensional). This situation is where the propensity score comes in. The **propensity score** represents the probability that a unit will belong to the treatment group and is based on pretreatment variables in a dataset. In other words, propensity scores can be used as a single-value summary of multiple variables. These scores can then be used to match treatment and control units, just like we have been doing with intercultural competence in our running example.

---

[4] Sometimes, researchers are especially interested in estimating the impact of the treatment on the treated units (the average treatment effect on the treated [ATT]) rather than the average treatment effect for all units in a dataset (ATE). In this case, no treatment units are dropped from the dataset, even though some control units are.

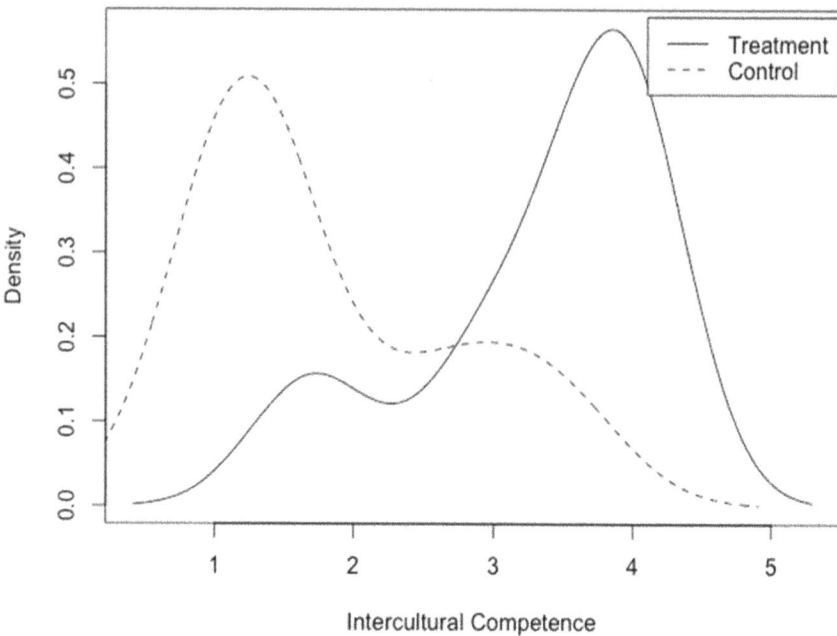

Fig. 9.1 Common support for running example (*Note* The data in this figure are invented for illustrative purposes.)

9.2.2 Example from the Literature

Iriondo's (2020) study is an example of how propensity scores can be used to match study participants when multiple pre-treatment variables predict whether a participant is part of the treatment or the control group. The purpose of Iriondo's (2020) study was to examine the relationship between participation in Erasmus study mobility (Erasmus is a student exchange program in the European Union) and salary and employment outcomes for recent college graduates in Spain. In this study, treatment students participated in an Erasmus mobility program while control students did not. The data that Iriondo (2020) uses in this study come from the Spanish National Institute of Statistics and contain a number of pre-treatment variables, including participant demographics, financial aid received while in school, and field of study, that were clearly associated with both participation in the Erasmus program and the two outcomes of interest: monthly salary and whether a student found employment after graduation. This is a situation wherein the threat that selection bias poses to the estimation of treatment effects is especially acute.

As is common in propensity score approaches, Iriondo (2020)'s analysis took place in two steps. In the first step, the author used a probit regression model to estimate the probability that a student would participate in an Erasmus mobility program. This probability became a student's propensity score. For most practical purposes, a probit regression model is the same as a logistic regression model,

Table 9.5 Mean values of covariates for treatment and control groups before and after nearest neighbor matching

Covariate	Treatment	Control (before matching)	Control (after matching)
Female	0.74	0.66	0.83
Academic records	0.08	0.00*	0.11
Father higher education	0.60	0.34***	0.62
Father primary	0.12	0.26***	0.12

(Adapted from Table 4 in Iriondo [2020])
*$p < 0.05$, ***$p < 0.001$

which we explored in Chap. 8. Once these propensity scores were estimated, Iriondo (2020) used nearest neighbor matching to match treatment and control students based on this score. Nearest neighbor matching is one of many possible matching approaches that researchers use when they carry out propensity score matching.[5] This matching approach, as its name suggests, involves matching each treatment participant to the control participant with the closest propensity score. For example, if a treatment student's propensity score is 0.942, then this student might be matched with a control student whose propensity score is 0.941 if this is the control group student with the closest propensity score.

An important component of this first step of propensity score analysis is to check the balance of the treatment and control groups before and after matching. In Table 9.2,[6] I show a subset of Iriondo (2020)'s balance check, which corresponds to participants' sex, academic record (that is, a measure academic achievement), and parental education level. The star levels in this table represent the results of a series of t-tests comparing means for treatment and control groups. As you can see, the treatment and control groups were significantly different from one another regarding academic achievement and parental education before matching. These significant differences were not present in the matched data (Table 9.5).

Once Iriondo (2020) obtained this matched sample, the second step of analysis involved using this new sample to estimate the relationship between Erasmus participation and the outcomes of interest: employment after graduation and salary. This second step of propensity score analysis corresponds to the standard regression analyses that were the topics of Chaps. 7 and 8, namely OLS or logistic regression. In Iriondo (2020), regression models using the matched data suggested

---

[5] A discussion of propensity score matching approaches is beyond the scope of this book, but several of the suggested readings at the end of this chapter delve deeper into matching techniques and their benefits and drawbacks.

[6] Note that Iriondo (2020) estimates the impact of Erasmus participation on employment and salary using two different datasets. The results presented in this section correspond to his results for the Labor Insertion Survey.

no significant impact of Erasmus participation on being employed or on salary in the time period immediately following a student's graduation. However, results did indicate a significant impact of Erasmus participation six years after graduation. Specifically, compared to non-participants, Erasmus participants had an 11 percent higher probability of being employed ($p < 0.05$) and earned salaries that were around 13 percent higher ($p < 0.05$).

9.3 Quasi-Experimental Design

While propensity score modeling can sometimes reduce bias in estimates of a treatment effect, this analytic approach relies on variables that are observed in a dataset, as noted previously. Propensity score modeling is unable to account for bias introduced into an analysis due to unobserved variables. To close out this chapter, I turn to two examples of **quasi-experimental design**, regression discontinuity and difference-in-differences.[7] In some contexts, and under specific assumptions, these two research designs are able to approximate the results that we would otherwise get from a true experiment, an RCT. Indeed, quasi-experimental design is the closest we can get to experimental design using observational data.

One difference between regression discontinuity and difference-in-differences approaches is the context needed for each design. Both approaches require that we know something about the way units are assigned to treatment and control groups, but this assignment mechanism is different for each approach. **Regression discontinuity** requires a situation wherein the way participants are assigned to treatment and control groups is standardized, such as through a placement score that determines whether students participate in a particular intervention. Students on either side of the cutoff placement score are thought to be as good as randomly assigned to treatment and control groups. **Difference-in-differences analysis** requires the presence of a specific policy or practice that takes place at a certain point in time and that sorts participants into natural treatment and control groups. An example of a policy that sorts participants might be a new initiative that requires students in certain degree programs to fulfill multicultural course requirements. Students in these specific degree programs would be the treatment group while those in other degree programs would be the control group. The following sections expand on regression discontinuity and difference-in-differences approaches in more depth.

---

[7] Regression discontinuity and difference-in-differences are not the only two quasi-experimental designs available to researchers (other quasi-experimental designs include instrumental variable and time-series analyses). However, they are two of the more common designs and are also possibly the most useful to individuals conducting research in international education.

9.3.1 Regression Discontinuity

As just described, regression discontinuity requires a situation wherein participants are assigned to treatment and control groups in a way that is known to the researcher. More specifically, in a standard regression discontinuity design, a continuous variable, such as a placement test score or ranking on a scale, determines treatment assignment. This variable is often referred to as a *running variable* or *forcing variable*. In most cases, regression discontinuity focuses analysis on observations on either side of a *cut point*, the point at which assignment into treatment or control groups is made.

To make these ideas more concrete, let us consider a situation where we want to know whether an academic scholarship that an institution awards to international students increases students' academic performance during the first term that they enroll. At this institution, once students enroll, they take an academic achievement test that is scored on a scale from 0 to 100. This score is the running variable. Students with a score of 50 or above are automatically awarded the scholarship while those with lower scores do not receive any funding. That is, the cut point for assignment to treatment and control groups is 50. At the end of the term, students' academic performance is measured again, also on a scale from 0 to 100. This measure is our outcome of interest. This situation is illustrated in Fig. 9.2, where students in the treatment group, those receiving the scholarship, are represented with black circles and students in the control group, those who do not receive the scholarship, are represented with grey circles. The placement score along the horizontal axis determines whether students receive a scholarship or not. This figure indicates a visible jump in students' end-of-semester academic performance (the vertical axis) around the cut point (at a score of 50), suggesting a detectable treatment effect for the scholarship (the significance of this treatment effect can be tested formally through a regression analysis where the predictor of interest is the treatment indicator).

Regression discontinuity makes the key assumption that around the cut point, students are as good as randomly assigned to treatment and control conditions. That is, while treatment and control students in this example on average may have very different characteristics regarding their home countries, native languages, prior academic achievement, and preparation to take standardized tests, factors that could influence both whether they end up in the treatment or control group *and* their future academic performance, those on either side of the cut point are very similar in terms of both observed and *unobserved* characteristics. Regression discontinuity often limits the analytic sample to individuals who fall within a certain *bandwidth* around the cut point. In our example, this bandwidth might limit the sample to students who scored within ten points of the cut point, between 40 and 60 on the placement test. It is good practice to check the balance between treatment and control groups on observed characteristics once a bandwidth is selected. It is also common for researchers to test multiple bandwidths, to explore how sensitive results are to the selection of a particular bandwidth (e.g., in our example, we might also test a bandwidth within five points of the cut point, between 45

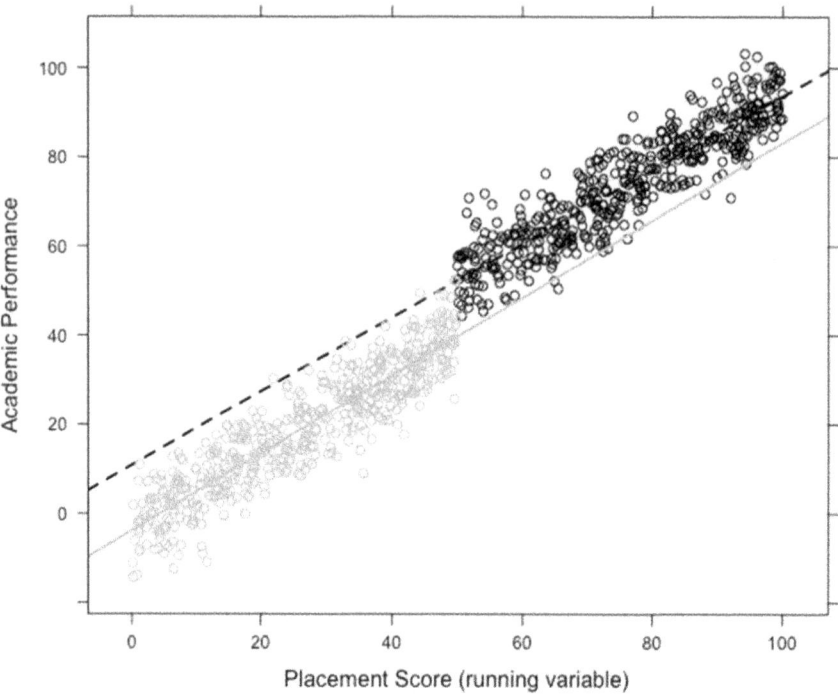

Fig. 9.2 Regression discontinuity design where treatment students scored over 50 on a placement test (*Note* The data in this figure are invented for illustrative purposes.)

and 55 on the placement test). The tradeoff is that a narrow bandwidth serves as a stronger control of outside factors aside from the treatment that might influence the outcome but risks having few observations in the analytic window, while a wide bandwidth likely includes more observations but is more susceptible to confounding. Hence, checking a few possibilities is advisable.

Because regression discontinuity uses only a subset of a dataset, the results of a regression discontinuity analysis are said to be *local* (and the treatment effect that is measured is referred to as the *local average treatment effect* or *LATE*), meaning that they apply only to those students around the cut point. An important assumption of regression discontinuity is that participants cannot manipulate the running variable. In our example, students have a motivation to manipulate their placement score so that it is higher, as higher-scoring students are the ones that receive scholarships. Such a situation could arise if students are allowed to take the placement test more than once if they are unhappy with their scores or if extra credit of some kind is given on the test. Additionally, regression discontinuity requires that the sample of students from around the cut point be large enough for statistical analyses to be valid. That is, if only three or four students cluster around the cut point, a regression analysis using only data from these students is not possible.

9.3.1.1 Example from the Literature

Baruffaldi et al. (2020) provide an example of regression discontinuity in international education research. This study explores the impact of an international mobility grant program for researchers sponsored by the Swiss National Science Foundation on several key researcher outcomes: short and medium-long term mobility, quantity and quality of research production, the researcher's academic position over time, and changes in a scholar's research network and research trajectory. Researchers who receive this grant affiliate with a host institution abroad for a maximum of 36 months and, while participants are encouraged to return to Switzerland after their time abroad, they are not required to do so. Applications for grant funding are reviewed by two external reviewers, who assign a numerical score. The Swiss National Research Council discusses these reviews and then assigns each application a final numerical score. Once normalized, this final score ranges from −4 to 2. In Baruffaldi et al. (2020)'s study, this normalized numerical score is the running variable. Although the cut point in this study is not a strict one (some applications with low scores were still funded), applications that received a 0 or higher were significantly more likely to receive funding. In our discussion here, we make an assumption that Baruffaldi et al. also make in part of their study—that the Swiss National Science Foundation funded applications awarded a score of 0 or higher and did not fund applications with lower scores. Funded applicants comprise the treatment group while unfunded applicants are the control group.

To begin their analysis, Baruffaldi et al. (2020) first inspect their data to ensure that a visible discontinuity (a visual jump) is present for their outcomes of interest at the cut point (such as in Fig. 9.2). If such a jump is not visible, then it is likely that there is no impact of the grant on the outcome. They then check the balance between their treatment and control groups, that is, researchers on either side of the grant eligibility cut point. Baruffaldi et al. find that there are no significant differences between individuals in each group around the cut point in terms of demographics, previous professional and academic achievement, previous research productivity, doctoral and host institution rank, amount of funding requested, or destination of proposed research. This balance check suggests that researchers at either side of the cut point are as good as randomly assigned to receiving the grant.

The basic form of the equation that describes the regression discontinuity model that Baruffaldi et al. (2020) estimate using ordinary least squares is the following:

$$Y_i = \alpha_i + \beta I(Score_i > Threshold) + \gamma X_i + \varepsilon_i$$

Here, β is the coefficient of interest, which measures the impact of the grant program on the researchers' outcomes of interest. The term $Score_i > Threshold$ means that an application's score was above the cut point, or threshold, of 0. This regression also controls for a number of researcher characteristics such as demographics and previous professional achievement (X_i) (see Table 9.6, which lists them all). These control variables are important to isolate the impact of receiving grant funding and to make the numerical measurement of this impact more precise.

Table 9.6 Impact of an international mobility grant on a researcher's average journal impact factor five years after receiving the grant

	Coefficient	Standard error
I(Grade > Threshold)	0.661*	0.365
Foreign	−0.174	0.282
Female	−0.252	0.340
Age	−0.038	0.039
Publications	0.044**	0.019
Citations	0.051***	0.011
Average journal impact factor	0.338***	0.077
Coauthors	−0.002	0.002
Early mobility grants	0.731**	0.333
Rank PhD university	−0.016	0.011
Rank host university	−0.017*	0.009
Rank host res. university	−0.022	0.017
Proposed duration	0.094	0.050
Amount request	−0.015	0.009
N	280	

(adapted from Baruffaldi et al. [2020] Table 4)
Note *** $p < 0.01$, ** $p < 0.05$, * $p < 0.10$. This regression model also includes destination dummies, commission dummies (corresponding to the Swiss commission evaluating the grant applications), and cohort fixed effects (N = 61)

Here, I focus on one of Baruffaldi et al.'s outcomes of interest: the average journal impact factor of a researcher's publications five years after applying for the grant. This average is taken to be an indicator of the quality of a researcher's publications. The results of this regression discontinuity analysis are summarized in Table 9.6. This analysis uses data from researchers whose applications received a score around the cut point of 0 (N = 280)—that is, these applications just barely made the cutoff to receive funding or were just barely below this cutoff and did not receive funding. As the result in the first row of this table indicates, applicants who received the grant had an average journal impact factor 0.661 points higher compared to applicants that did not receive the grant ($p < 0.10$). The authors conclude that this result indicates that the international mobility grant is directly related to an increase in the quality of researcher productivity. Funding, it seems, has a direct and causal relationship with research quality (at least to the extent that impact factor is an accurate indicator of this outcome) (Table 9.6).

9.3.2 Difference-In-Differences

Difference-in-differences is a second quasi-experimental research design that is especially useful in international education research. This type of research design is

9.3 Quasi-Experimental Design

often used to explore the impact of policy changes or the introduction of new programs (I will use the term *initiative* to refer broadly to these two categories in this section). To conduct a difference-in-differences analysis, researchers need access to longitudinal data that follow units (students, institutions, etc.) over time both before and after an initiative's implementation. An additional requirement for conducting a successful difference-in-differences analysis is that natural treatment and control groups have to be present in the data. To illustrate difference-in-differences, let us consider an initiative wherein a research university begins to offer grant funding in two STEM fields, Engineering and Biology, to encourage increased international research productivity in these areas. Faculty members in Engineering and Biology comprise the treatment group while those in other STEM fields are the control group.

To begin our analysis, we can graph international research output, defined as the number of publications co-authored with an international researcher, over time for Engineering and Biology faculty. This graph appears in Fig. 9.3, where academic years, referred to as time periods, are along the horizontal axis and international publications are shown on the vertical axis. Knowing that the university began offering international research funding incentives in the sixth year captured in the dataset, we might take the jump in publications between the fifth and sixth years (and the sustained increase in research productivity in subsequent years) as evidence that the funding initiative served its purpose.

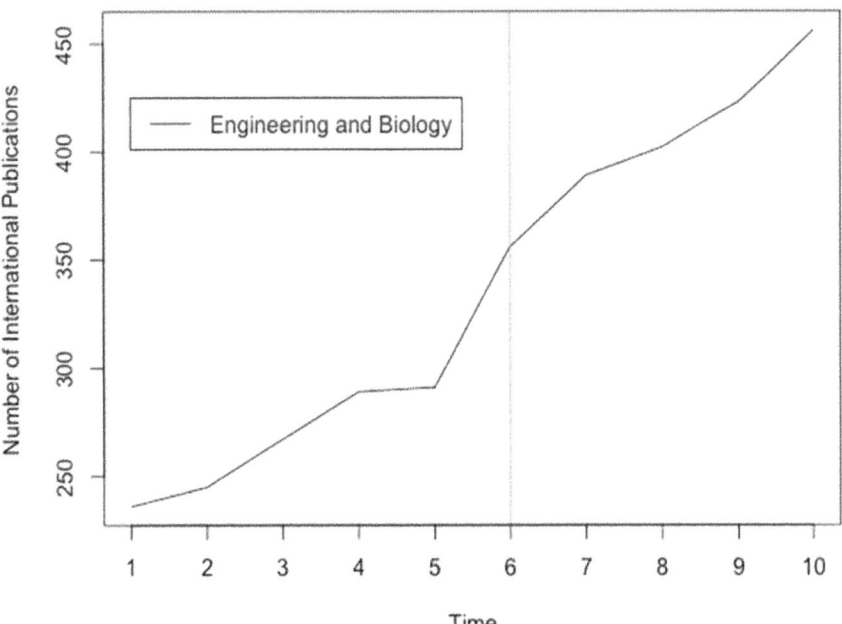

Fig. 9.3 Publications by engineering and biology faculty over time (*Note* The data in this figure are invented for illustrative purposes.)

However, notice that even in years before the initiative, the number of international publications from faculty in Engineering and Biology was increasing. How do we know that the increase in year six was due to the university's international research initiative and not some other factor that might have impacted all STEM faculty? For example, perhaps the Arts and Sciences unit at this university implemented new requirements for tenure and promotion, such as an increase in the number of expected publications, in this same year. These new tenure and promotion requirements could be responsible for the jump in international publications that we see in Fig. 9.3 and would correspond to an increase in international publications among faculty across STEM fields, even in the absence of the international research funding initiative.

To unravel the impact of the university's international research funding initiative for Engineering and Biology faculty from the impacts of other changes that might be responsible for an increase in international research publications over time, we need a comparison group that is similar to the treatment group but that would not have been impacted by the funding initiative. An obvious comparison group is faculty members in STEM fields that are not Engineering or Biology. Changes in tenure and promotion expectations, for example, would have impacted this group's publication patterns similarly to the Engineering/Biology faculty group, while the international research funding initiative would not. Figure 9.4 shows the time trend in international co-authored publications for both faculty groups.

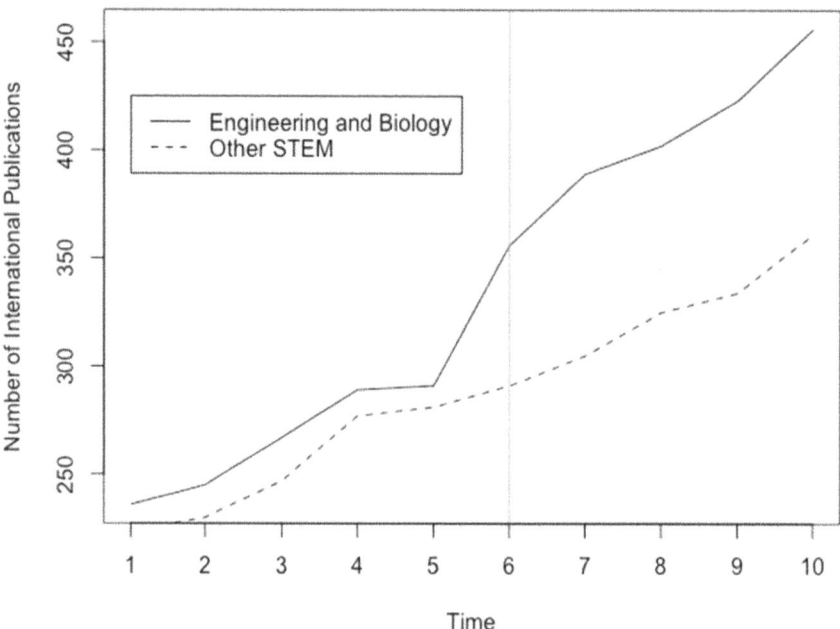

Fig. 9.4 Publications by engineering and biology faculty and faculty in other STEM fields over time (*Note* The data in this figure are invented for illustrative purposes.)

9.3 Quasi-Experimental Design

In Fig. 9.4, we see that even before the implementation of the international research initiative, faculty in the treatment group published with international co-authors more frequently than those in the control group. This difference is not necessarily surprising—we did not randomize faculty members into treatment and control groups, but rather used groups that occurred naturally. We would expect to observe some differences between these two groups. Perhaps Engineering and Biology as fields have a generally stronger culture of international collaboration compared to other STEM fields, or maybe faculty members in these two fields at this specific university attended graduate programs that happened to have strong international connections. Regardless, we also see in Fig. 9.4 that prior to the funding initiative, the time trends of the two faculty groups were very similar to one another, meaning that while those in the Other STEM group co-authored with international researchers less than those in Engineering and Biology, their publication trends were otherwise very similar. This similarity in time trends satisfies what is called the *parallel trends assumption* in difference-in-differences analysis and suggests that the Other STEM faculty group is a good comparison group for the treatment group. That is, once we account for factors explaining the differences in international research publication output between the two groups in the pre-treatment time period, the two groups follow very similar publication patterns.

When conducting a difference-in-differences analysis, researchers take advantage of parallel trends to estimate what would have happened to the treatment group had the initiative of interest not been implemented. This alternate reality, often referred to as a *counterfactual*, is illustrated in Fig. 9.5. In this figure, what would have happened to the international publication trend for Engineering and Biology faculty had the initiative never happened is represented with the dotted (rather than the solid) line. The impact of the funding initiative is estimated by comparing this dotted line (the counterfactual) and the solid line representing what actually happened to international co-authored publications for the treatment group.

To measure the average distance between the dotted and solid lines (the counterfactual and actual number of publications for Engineering and Biology faculty) in Fig. 9.5 after the implementation of the funding initiative in year six, we need to make two different comparisons: a comparison over time for each experimental group individually (that is, a within-group comparisons) and a comparison of these differences between the two groups before and after the initiative (that is, a cross-group comparison). In its simplest form, represented in the following equation, a difference-in-differences estimate simply involves taking the differences (the cross-group comparison) between two differences (the within-group comparisons) (hence the name of the analytic approach, difference-in-differences). The first step in a basic difference-in-differences analysis involves subtracting the mean outcomes for the treatment and control groups before policy implementation (called the pre-treatment time period) from their corresponding means after policy implementation (called the post-treatment time period). In the following equation, these differences are represented as $Publications_{t1} - Publications_{t0}$ for the treatment group and $Publications_{c1} - Publications_{c0}$ for the control group, where t stands

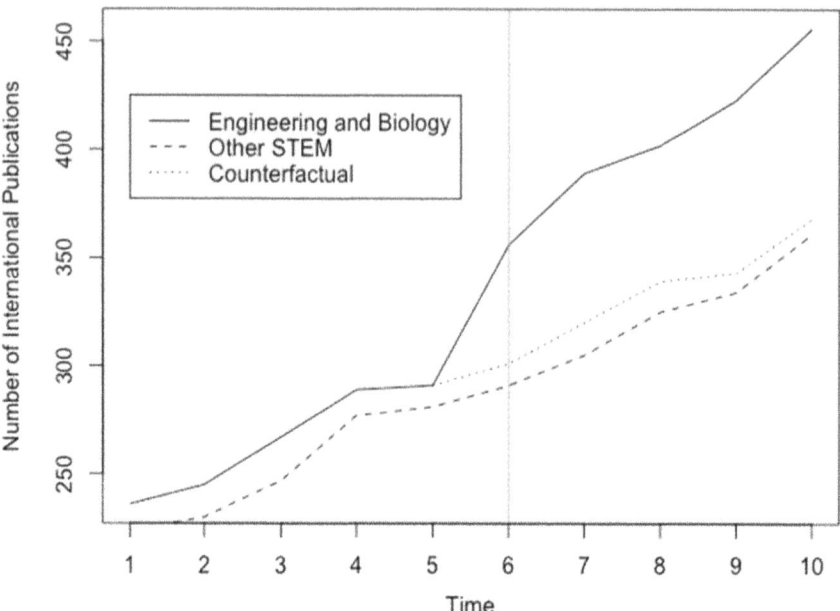

Fig. 9.5 Publications by engineering and biology faculty, faculty in other STEM fields, and the counterfactual scenario over time (*Note* The data in this figure are invented for illustrative purposes.)

for *treatment*, *c* stands for *control*, a 1 represents the post-treatment time period, and a 0 represents the pre-treatment time period. Next, we subtract these two differences from one another to calculate the estimated impact of the initiative.

$$\text{Initiative Impact} = (Publications_{t1} - Publications_{t0})$$
$$- (Publications_{c1} - Publications_{c0})$$

In our example, the mean number of internationally co-authored publications for engineering and biology faculty before the implementation of the funding initiative was 265.6. This mean jumped to 405.2 after the funding initiative. Meanwhile, for faculty in other STEM fields, the mean number of internationally co-authored publications was 251.8 before the policy initiative and 323.2 afterwards. If we plug these numbers into our difference-in-differences equation, we get the following result:

$$\text{Policy Impact} = (405.2_{t1} - 265.6_{t0}) - (323.2_{c1} - 251.8_{c0}) = 68.2$$

That is, our estimated impact of the university's international research funding initiative for faculty in Engineering and Biology was an average increase in around 68 internationally co-authored publications.

9.3.2.1 Example from the Literature

It is possible to extend the logic of a difference-in-differences analysis to a regression context to control for other variables that predict a given outcome variable. In our running example, we might think that other factors, such as faculty demographics or other grant funding they receive, would impact how frequently faculty members co-author with international collaborators. We would want to include these variables in our statistical model. Regression also provides us with a tool to test whether our detected treatment effect (68.2 in this example) is significantly different from zero.

To illustrate how to incorporate covariates into a difference-in-differences analysis, I summarize a study I conducted that examines the impact of a certain kind of financial aid program, merit aid programs (the treatment), on a U.S. state's study abroad participation rate (the outcome) (Whatley, 2019). In the United States, individual states often implement these merit-based financial aid programs, which provide scholarship funds to students with a certain level of academic achievement, for the purpose of retaining high-achieving students within state borders. Some states have merit aid programs and some states do not. While I expected that a policy that provided students more educational funding would lead to an increase in a state's study abroad participation rate (more students could afford to study abroad), merit aid programs also often require students to remain on campus for a certain part of their degree programs or have other requirements that are not logistically compatible with study abroad participation. In this case, a negative impact of merit-aid initiatives on study abroad participation is also possible. This study used a dataset that spanned the years 1988 through 2014 and included information about states' study abroad participation rates (taken from the Institute of International Education's Open Doors reports) and other state characteristics that I thought might impact study abroad participation (e.g., one of these variables was the state's unemployment rate, which I took from the U.S. Bureau of Labor Statistics). These state characteristics were my control variables in my regression-based difference-in-differences analysis.

Since study abroad participation rate is a continuous variable, I used ordinary least squares regression. My regression took the following basic form:

$$Study\ Abroad\ Rate = \alpha + \beta_1 Treat + \beta_2 Time + \beta_3 (Treat * Time) \\ + Controls\gamma + \varepsilon$$

In this equation, the outcome of interest is a state's study abroad participation rate (*Study Abroad Rate*) while *Treat* is a binary indicator equal to 1 if a state had a merit aid program and equal to 0 otherwise (1 = treatment, 0 = control). *Time* is a binary indicator of time period (1 = post-policy implementation, 0 = pre-policy implementation). The variable *Treat * Time* represents these two indicators multiplied by one another, meaning that this interaction term equals 1 only for treatment states in the post-policy implementation period, the time during which treatment states would have received treatment. The coefficient β_3 is the difference-in-differences estimate of the policy impact. In the absence of any control variables, this coefficient equals the naïve difference-in-differences treatment estimate that I would obtain by subtracting the pre- and post-policy outcomes for treatment and control groups like I did previously. To account for control variables in my regression model, I simply entered them into my analysis (*Controls*γ), just like I would in a normal ordinary least squares regression.

Table 9.7 summarizes my results for this analysis, including control variables. My results suggested that the implementation of a merit aid program had a negative

Table 9.7 Difference-in-differences estimates of the impact of state merit aid programs on the state's study abroad participation rate

	Regression coefficient	Standard error
Difference-in-differences estimate (Time*Treat)	−0.317***	0.082
Time	0.196**	0.069
Treatment	−0.016	0.072
Unemployment rate	−0.047***	0.010
Median household income[a]	−0.000	0.000
% Bachelor's degree or higher	0.014**	0.005
State appropriations to higher education[ab]	−0.000*	0.000
Total state need-based grant aid to	0.000	0.000
Total state non-grant aid to colleges/universities[ab]	−0.002***	0.000
Total state non-need-based grant aid to colleges/universities[ab]	0.001***	0.000
Average public tuition[ac]	0.140***	0.012
Average private tuition[ac]	0.027***	0.004
Proportion of students in liberal arts	0.401	0.427
Proportion of racial/ethnic minoritized students	−0.121+	0.064
Proportion of female students	0.035	0.192
(Intercept)	−0.127	0.167
Sample Size	817	
R^2	0.62	

(Reproduced from Table 5 in Whatley [2019])
Standard errors in parentheses, +$p < 0.10$, * $p < 0.05$, ** $p < 0.01$, *** $p < 0.001$
[a] Adjusted to 2014 dollars
[b] In millions
[c] In thousands

impact on a state's study abroad participation rate. Specifically, once I accounted for control variables, establishing a merit-aid program was estimated to decrease a state's study abroad participation rate by around 0.32 percentage points. While this decrease seems small, it was statistically significant ($p < 0.001$).

9.4 Future Quantitative Research in International Education

The purpose of this chapter was to introduce readers to experimental and quasi-experimental research design. These analytic approaches allow for valid and robust estimates of treatment effects, a very powerful use of quantitative analysis. This is not to say that the descriptive approaches that comprise the majority of this book are unimportant. Sometimes, the purpose of a research project is to describe. These descriptive designs also form the basis of all the research designs described in this chapter, and a key first step in any quantitative research study is to describe the tendencies present in data. At the same time, experimental and, more practically, quasi-experimental research designs are likely the future of quantitative analysis in international education research. These approaches to research do not ignore the fact that in international education, we study social actors (whether humans or organizations and other entities that humans create) that make their own decisions about international education-focused policies and practices. The ability of experimental and quasi-experimental designs to estimate accurate treatment effects in these non-experimental contexts is essential to advancing international education theory and producing accurate statistical analyses to inform policy and practice.

Recommended Reading

A Deeper Dive

Cunningham, S. (2021). *Causal inference: The mixtape*. Yale University Press.

DesJardins, S. L., & Flaster, A. (2013). Nonexperimental designs and causal analyses of college access, persistence, and completion. In L. W. Perna & A. Jones (Eds.), *The state of college access and completion: Improving college success for students from underrepresented groups* (pp. 190–207). Routledge.

Diamond, A., & Sekhon, J. S. (2013). Genetic matching for estimating causal effects: A general multivariate matching method for achieving balance in observational studies. *Review of Economics and Statistics, 95*(3), 932–945.

Furquim, F., Corral, D., & Hillman, N. (2020). A primer for interpreting and designing difference-in-difference studies in higher education research. In L. Perna (Ed.), *Higher education: Handbook of theory and research* (Vol. 35, pp. 2–53).

Iacus, S. M., King, G., & Porro, G. (2011). Multivariate matching methods that are monotonic imbalance bounding. *Journal of the American Statistical Association, 106*(493), 345–361.

Murnane, R. J., & Willett, J. B. (2011). *Methods matter: Improving causal inference in educational and social science research*. Oxford University Press.

Reynolds, C. L., & DesJardins, S. L. (2009). The use of matching methods in higher education research: Answering whether attendance at a 2-year institution results in differences in educational attainment. In *Higher education: Handbook of theory and research* (pp. 47–97). Springer.

Rosenbaum, P. R., & Rubin, D. B. (1985). Constructing a control group using multivariate matched sampling methods that incorporate the propensity score. *The American Statistician, 39*(1), 33–38.

Rubin, D. B. (1973). The use of matched sampling and regression adjustment to remove bias in observational studies. *Biometrics*, (Vol. 29, pp.185–203).

Additional Examples

d'Hombres, B., & Schnepf, S. V. (2021). International mobility of students in Italy and the UK: Does it pay off and for whom? *Higher Education.* https://doi.org/10.1007/s10734-020-00631-1

Dicks, A., & Lancee, B. (2018). Double disadvantage in school? Children of immigrants and the relative age effect: A regression discontinuity design based on the month of birth. *European Sociological Review, 34*(3), 319–333.

Marini, G., & Yang, L. (2021). Globally bred Chinese talents returning home: An analysis of a reverse brain-drain flagship policy. *Science and Public Policy.* https://doi.org/10.1093/scipol/scab021

Monogan, J. E., & Doctor, A. C. (2017). Immigration politics and partisan realignment: California, Texas, and the 1994 election. *State Politics & Policy Quarterly, 17*(1), 3–23.

10. Writing About Quantitative Research

The purpose of this chapter is to explore how to communicate the results of quantitative research. While this is the last chapter of the book, it is perhaps the most rather than the least important. After all, we conduct research to communicate about our findings to others. In the first section of this chapter, I discuss writing for scholarly audiences, which are often the readers of course term papers, academic assignments, research-focused conference proposals, and academic manuscripts. For better or for worse, when writing for an academic audience, there is an expected way in which information is presented to readers in quantitative research. In this chapter, I detail the outline that I always begin with when I write for an academic audience and that I share with students when requiring them to write about their own research for final assignments in the courses I teach. The second section of this chapter focuses on writing about quantitative research for external audiences. In my experience, these external audiences include primarily policymakers and practitioners in the field. These are often audiences to whom we want to communicate our research so that they can put it to use in revising existing policy, crafting new policy, or changing and creating practices in their professional environments.

10.1 Scholarly Writing

While the focus of this book is centered on the analysis of quantitative data, the framing of quantitative research—for all audiences, not just academic ones—involves more than just data and analysis. As quantitative researchers, we often get caught up in the details of an analysis and forget to take a step back and look at the bigger picture of what drives the methodological decisions and assumptions we make. This bigger picture becomes especially important when we sit down to write about our quantitative work (a list of current journals that publish international education research can be found in Appendix D). Written pieces about quantitative research involve situating the study in previous literature on the topic, providing theoretical foundations for the study, and returning to previous literature and theory when discussing and interpreting results. To effectively make this connection between the beginning and the end of a written scholarly work, prior literature and theory must be effectively threaded through the analytic approach that a researcher uses. When writing about your work, these connections need to be made clear.

The outline that I begin with when I write about a study that I have conducted almost always looks like this:

1. Introduction
2. Theoretical Framework
3. Literature Review
4. Method
   a. Data
   b. Analysis
5. Results
   a. Descriptive Statistics
   b. Primary Statistical Models
6. Discussion
7. Conclusion.

Here, I expand a bit on each of these sections.

10.1.1 Introduction

The introduction of an academic paper describes the problem or gap in the literature that the study addresses, identifies the research questions or hypotheses that the study informs, explains briefly how these questions or hypotheses are approached, both theoretically and methodologically, and explains why the study is important.

A common way to begin a quantitative research paper is to provide some information about the overarching context of the study. For example, in their study about international students' friendship networks (summarized in Chap. 6: Correlation), Hendrickson et al. (2011, p. 281) begin as follows: "International students constitute a diverse and unique group of students at universities in the United States and countries across the globe. Overseas students are one of the most intensely studied groups in the culture contact literature (Ward et al., 2001)." Iriondo, in his study of the impact of Erasmus study mobility on graduates' salaries and employment in Spain (summarized in Chap. 9: Introduction to Experimental and Quasi-experimental Design), introduces his study providing some overarching statistics on European and Spanish participation in the Erasmus program: "Since its implementation in 1987, over 3 million European students have participated in the Erasmus study mobility program. Over time, Spain has established itself, in terms of number of participants, and the program's main country of origin and destination" (Iriondo, 2020, p. 925).

A key component of the introduction of a quantitative research paper is a purpose statement, which describes exactly why the research was conducted. Examples of purpose statements, taken from Hendrickson et al. (Example A) and Iriondo (Example B), respectively include:

> **A:** "The intended research goals of this study are to examine how international students manage their social resources; i.e. friendship networks, using network theory principles, and identify patterns that aid international students in achieving greater satisfaction and success with their study abroad experience. The pragmatic goal lies in the hope that this knowledge may help create a deeper cultural understanding between the host country and the international student." (Hendrickson et al., 2011, p. 282)
>
> **B:** "The main objective of this article is to evaluate the impact of the Erasmus program on graduate salary and employment prospects in Spain. Despite the scale of the Erasmus program, the availability of empirical studies that evaluate its impact on the Spanish labor market are scarce. Moreover, given that participants in mobility programs present different characteristics from their peers in terms of ability, field of study or socioeconomic background, it cannot be stated with any certainty that the correlations observed to date are in fact causal." (Iriondo, 2020, p. 927)

These statements provide not only information about what each of these studies accomplishes (insight into international students' social networks and an evaluation of the impact of the Erasmus program on salaries and employment, respectively), but also, in the case of Hendrickson et al., insight into the theoretical framework that informs the study (network theory) and, in the case of Iriondo, hints about the study's methodological approach (causal inference). These statements also provide some information about why the studies are important. Hendrickson et al. mention the importance of their findings to the creation of deeper cultural understanding between international students and their host countries, while Iriondo mentions the scale of the Erasmus program, suggesting that considerable financial resources are used to fund a program whose outcomes are not at all clear.

Research questions or hypotheses often derive directly from purpose statements. Note that while I tend to recommend that research questions and/or hypotheses come at the end of the Introduction section, these components of a study might make more sense after the Theoretical Framework and Literature Review sections, particularly if they require substantial background knowledge or key terminology from the empirical or theoretical literature. Iriondo's study provides two research questions at the beginning of the Methods section, after the logic behind propensity score modeling has been outlined:

> "(1) Does participation in an Erasmus mobility program increase the employability of recent graduates? Erasmus mobility may increase the likelihood of finding employment, given that experience in an international environment provides skills that are valued positively from the perspective of employers. (2) Does the experience gained from Erasmus mobility bring about human capital enhancement and a corresponding increase in salaries? International

student mobility often leads to a greater mastery of foreign languages, possibly impacting productivity and salaries positively." (Iriondo, 2020, p. 930)

Notice that in this example, Iriondo outlines research questions that essentially correspond to one-tailed hypotheses—that is, he hypothesizes a positive effect of the Erasmus program on graduates' employment and salary outcomes. Moreover, Iriondo provides the reason why he expects the Erasmus program to have a positive impact on outcomes—the program provides students with skills, particularly foreign language skills, that employers value. These hypotheses related directly to Iriondo's theoretical approach, human capital theory, which suggests that individual investment in education will be rewarded in the labor market (Becker, 1994).

10.1.2 Theoretical Framework and Literature Review

The theoretical framework and literature review sections of a quantitative research paper often expand on concepts and ideas mentioned in brief in the Introduction section, and both sections serve to provide a rationale for research questions and hypotheses and methodological decisions that the researcher makes. Regarding theory, we just saw an example of how Iriondo's (2020) research questions draw from human capital theory. That is, human capital theory predicts that personal investment in educational activities that enhance an individual's human capital will in turn be rewarded in the labor market. In Iriondo's study, investment in educational activities involves participation in the Erasmus program, and rewards in the labor market include employment and salary. Hendrickson et al. (2011) outline a functional theoretical model (developed by Bochner et al., 1977) that describes the friendship formation of international students. This model classifies international students' friendships into three categories: co-national friendships, host national friendships, and multi-national friendships, each of which serve a separate function for international students. Co-national friendships affirm a student's culture of origin, host national friendships facilitate a student's aspirations in the host environment, and multi-national friendships serve a recreational purpose. This three-group categorization of friendships is reflected in Hendrickson's methodological approach, which separates out friendships into these three categories for both data collection and subsequent analysis (recall from Chap. 6 that, in this study, correlations between strength of international students' friendships and measures of satisfaction, contentment, homesickness, and connectedness were conducted separately for co-national, host national, and multi-national friendships).

Previous literature can also help researchers form hypotheses and research questions and think critically about methodological decisions. For example, if previous research has found a positive association between human capital investment and employment outcomes, it might make sense to put forth a similar hypothesis in your own research. If previous research has found differences in international mobility participation among students belonging to different demographic groups, then you might expect to see a similar pattern in your own data. Methodologically, previous literature can also inform who your population of interest is or which variables to include in your analyses. For example, if all the previous research that explores the relationship between international student mobility and the development of intercultural competence has focused on students attending four-year institutions, then you might use this fact to justify why your study of the same issue among students attending a different kind of institution is important. Regarding variable selection, if your aim is to isolate the impact of a particular program on student outcomes, like Iriondo does in exploring the impact of the Erasmus program on students' employment outcomes, then you likely need to include control variables in your analyses that correspond to other variables, identified in previous literature, that relate to the same outcomes. For example, Iriondo highlights especially how academic performance, regardless of whether a student participates in international mobility opportunities, has been shown in previous literature to predict subsequent employment success. For this reason, he includes a measure of students' academic performance as a control variable in his analyses. Important to note is that the purpose of a literature review in an academic paper is not to summarize all the literature that has ever been written about a specific topic. As described in this section, the purpose is to discuss and provide examples of previous literature that inform different aspects of your study—such as the research questions and/or hypotheses or particular methodological decisions.

10.1.3 Method

In general, the Method section of a quantitative academic paper contains two primary sections: Data and Analysis. The overall goal of the method section is to communicate clearly and efficiently the actions that you took to conduct the study. A useful way to think about this section of a paper is that a reader should, at least hypothetically, be able to read this section and recreate the same study that you conducted. Important to note here is that these steps are inevitably influenced by your own perspective of the world and that the actions you take rest in part on your own background and experiences. While including explicit mention of your positionality as a researcher is not commonplace in published quantitative research, writing this section in first person or providing some other explicit mention of your personal connection to the research project can help readers of your work understand the context within which you conducted your study and the perspective with which you view your study and, ultimately, interpret your findings.

A common situation is that we become interested in a particular research project because we ourselves are personally involved in its topic in some way. For example, I arrived at the research project that explored the relationship between merit-based financial aid and study abroad participation, outlined in Chap. 9, because I, myself, both received merit-based financial aid from one of the statewide programs that was the focus of the study (in my case, Georgia's HOPE scholarship) and participated in study abroad. As a student, I often heard students and administrators alike make the connection between merit-based financial aid and some of the decisions that students made, and I wanted to test out the hypothesis that such a connection existed for study abroad. In fact, in the original version of the study, my hypothesis was directional—I thought that receiving merit-based financial aid would increase the likelihood that students would study abroad. I arrived at this hypothesis in part based on my own experience. While traditional quantitative methods literature suggests that such research is unbiased, objective, and not at all influenced by the perspectives and experiences of the researcher (a philosophical approach often referred to as *positivism*), my example suggests that this is not always the case. In fact, I suspect that this is not *usually* the case and, as a consequence, I advocate wholeheartedly for spending more time and attention on our own positionalities when we conduct quantitative research.

Data. The data section of a quantitative academic paper provides information about the data, often from diverse sources, that inform the research questions or hypotheses posed previously. If you are using survey data, this is the section where you describe your survey—including information about how you designed it, what questions you asked, and who your participants were. Here, you would provide information such as example survey questions, connections between your survey questions and theory or previous research, and information about data collection (e.g., random sampling, convenience sampling) and participant response rate. If your study includes a measurement instrument, this is the section where you would describe the instrument. For example, Hendrickson et al. (2011) used three different measurement instruments in their study: a social connectedness scale; a homesickness and contentment scale; and a satisfaction with life scale. In the section where they describe their data, they provide information about these scales, including a brief description as well as information about the scale's reliability and validity, which speak to the quality of the scale.

If you are using secondary data (e.g., from an external data source), the Data section is where you describe the key variables you used in your study and justification for why each one was included, again, based in previous literature or theory. Iriondo (2020) is an example of a study that relies on secondary data. In his data section, Iriondo provides an overview of his data source for readers who are unfamiliar, the Spanish National Institute of Statistics, and describes the information about students in each of the two datasets that he uses in his study. Other important information to include in this section might be the time period when data were collected, detailed descriptions of key outcome and predictor variables, especially if these are variables researcher-derived (e.g., Iriondo's outcome for income is the logged value of this variable).[1]

Regardless of the data source, whether you collected your own data or are using secondary data, this subsection of your Methods section should include information about how many observations you include in your analyses, if and why you dropped particular observations, and how you dealt with missing data. This last

---

[1] A full treatment of when and why researchers log variables is beyond the scope of this book, but interested readers are referred to Gujarati and Porter, 2010, Chap. 4 for additional information.

point on missing data is important as simply excluding observations with missing data can greatly bias your results. For example, if your population of interest comprises all U.S. institutions of higher education and one of your key variables of interest is an institution's acceptance rate, some institutions will be missing data on acceptance rate because they do not have one—by definition, open-access institutions admit all students who apply. In other words, dropping institutions with missing acceptance rates will systematically exclude a certain type of institution from your analyses. As a researcher, you will need to make important decisions about how to address missing data so that your results are not biased.[2]

Analysis. Similar to the Data sub-section, the Analysis sub-section should provide clear and detailed information about how you analyzed the data in your study. You should provide sufficient information so that if someone were to want to replicate your study on their own, they could. In many ways, the chapters that comprise this book provide a blueprint for what information to include in your analysis section, which of course depends on the kind of analysis you use. For example, if you are conducting an ordinary least squares regression analysis, you will need to include information about your outcome variable and predictors of interest, as well as information about specific control variables. If you include interaction terms in your regression model, then you would describe these variables. Similarly, if your study used propensity score modeling, you would detail the two phases of this analytic approach, describing first how you constructed your propensity scores and second how you constructed the model predicting your outcome of interest. Importantly, in this section, you should justify why you selected the analytic approach that you chose. For example, if you choose to analyze your data using *t*-tests rather than a regression analysis, you will want to explain why you do not need to include control variables in your analysis, as would be the case in a regression.

Equations. The Analysis section is the most likely place for you to include equations to describe for your reader exactly how you constructed a statistical model. It is important that, in addition to providing the equation for a particular analysis, you also describe the different pieces of the equation for your readers. Iriondo (2020) provides an example of this practice[3]:

"As a means of evaluating the effect of Erasmus mobility on employment and salaries, estimations of the following linear regression models were made:

$$E_i = \beta_1 + \beta_2 Erasmus_i + \beta_3 X_i + \beta_4 F_i + \beta_5 R_i + \varepsilon_i \qquad (10.1)$$

$$\log w_i = \gamma_1 + \gamma_2 Erasmus_i + \gamma_3 X_i + \gamma_4 F_i + \gamma_5 R_i + u_i \qquad (10.2)$$

---

[2] While a full discussion on different approaches to handling missing data in quantitative research is beyond the scope of this book, interested readers can refer to Cox et al. (2014) (Working with missing data in higher education research: A primer and real-world example) and Manly and Wells (2015) (Reporting the use of multiple imputation for missing data in higher education research) for in-depth discussions of some of the more common ways to address missing data.

[3] See Footnote 1 in this chapter regarding logged outcome variables.

where E stands for employment and it is a dichotomous variable that takes value 1 when the graduate was in employment and 0 otherwise. $\log w$ is the natural logarithm of the graduate income, w. *Erasmus* is a dichotomous variable that takes value 1 if the graduate participated in the mobility program and 0 otherwise. X is a vector of graduate training and personal attributes that bear an influence on earnings and on the probability of obtaining employment [...]. Finally, F represent a set of dummy variables related to the field of study and R, a set of dummy variables related to the region where study took place." (Iriondo, 2020, p. 930).

Notice that the text provided by Iriondo surrounding Eqs. (10.1) and (10.2) help clarify for readers exactly what each piece of the equation means.

When including an equation in your writing, two additional pieces of advice also apply. First, use the equation function in your word processing program to write out the equation rather than try to find the appropriate symbols and other notation as regular type. Equations entered using tools designed specifically to accommodate mathematical symbols are clearer and easier to read. They also allow you to use proper notation, such as Greek letters or subscripts, in a simple format. Second, wedging words into equations (as Iriondo does with the word "Erasmus") may work with short words or short equations, but is likely not the optimal choice when words or equations are longer. In these cases, using letters or abbreviations to indicate variables and then explaining further in your text is, again, a clearer and easier-to-read approach.

10.1.4 Results

In a standard format, the results of your quantitative analyses are provided *after* you describe your research methods. I typically start with two sub-sections of my Results section (and also recommend this format to students) that I can then expand as needed. The first of these sections provides descriptive statistics while the second provides the results of my primary statistical models, which are usually regression analyses of some kind.

Descriptive Statistics. Descriptive statistics include measures of central tendency (Chap. 2) and measures of variability (Chap. 3) for the variables included in a study's analyses. While it is relatively standard to include this information for *all* the variables you use in your study (including control variables if you are using regression analysis), additional details are often provided for key variables of interest. The rationale for including this information for all variables in a study is twofold. First, information about the means and standard deviations for all variables can provide readers with a general idea of what the average unit in your data is like. Second, this information provides insight into the quality of your data. That is, these statistics provide both you and your readers with the opportunity to determine if certain descriptive information is realistic (e.g., if the average student in your dataset is recorded as taking 12 courses per term, then one might question the accuracy of this variable). Additional details about key variables of interest in your study can also be informative. For example, if your study's goal is (in part) to analyze differences between students who study abroad and those who do not, in addition to providing descriptive statistics for your overall sample, it would likely be useful to also provide this information for samples of study abroad participants and non-participants separately.

Primary Statistical Models. After presenting descriptive information about your study's sample, the next piece of a Results section contains the results of the study's primary statistical models. In my work, this usually means that this section outlines the results of a series of regression analyses (keeping in mind that all the analytic approaches described in Chaps. 7, 8, and 9 are regression-based), but this section might also outline the results of other analyses, such as ANOVA or Chi-square tests. When writing about specific results, it is common practice, illustrated throughout the chapters in this book, to include the test statistics related to a particular result in parentheses at the end of a sentence. Hendrickson et al. (2011) provide a good example of this practice: "An independent samples t-test revealed no statistically significant effects between locals and non-residents on respondents' satisfaction ($t(72) = 1.64$), contentment ($t(71) = 1.12$), homesickness ($t(72) = -0.50$), or connectedness ($t(72) = 1.69$)" (pp. 287–288). Here, we see that the authors provide the calculated t-value for each of the t-tests that they conducted along with the associated degrees of freedom (in parentheses) for each test. If I wanted to explore the statistical significance of these calculated t-statistics myself using Appendix A, I could look up the critical t-values associated with the degrees of freedom that the authors provide.[4]

A Note on Tables and Figures. Results sections are often a common place where we use tables and figures to summarize a study's findings. Tables and figures provide an easily accessible and efficient way to display results—especially if analyses involve a large number of variables or if several similar analyses are conducted. Two key pieces of advice to keep in mind when using tables and figures in an academic paper are to ensure that tables and figures are interpretable on their own (that is, someone could look at the table or figure and understand what kind of analysis the researcher used, key information about the data that produced the results, and what the results themselves indicate) and to describe tables and figures in the text to direct readers' attention to important pieces of information. To illustrate this practice, I reproduce here as Table 10.1 a section of Table 7 from Iriondo (2020, p. 937).

Notice that when looking at Table 10.1, I know immediately that the results it displays are from an analysis using propensity score matching, as noted in the title of the table. I also know from the title that the treatment indicator in the study is Erasmus mobility and the outcome of interest is employment. The column titles tell me that the table provides results for employment after a student's graduation and six years later (presumably six years after graduation). The notes at the bottom

---

[4] Note that it is also good practice to report estimates and measures of uncertainty that go into any test statistic. These might not be written in the main results, but at least in the appendix it is a good idea to report such information. In a case like this, the sample means that were compared and the associated standard error used to calculate t should be available to the well-versed reader.

Table 10.1 PSM estimation of the impact on employment of Erasmus mobility (Iriondo, 2020)

Variables	After graduation	6 years later
Erasmus	0.040	0.106***
	(0.062)	(0.048)
Observations	142	192
R^2	0.201	0.254

Source Own calculations from EIL-2008 (UCM)
Notes Robust standard errors in brackets [$***p < 0.01$, $**p < 0.05$, $*p < 0.1$]. The following explanatory variables are included in the estimation of the employment equation: female, father's education, academic records, postgraduate education and field of study

of the table tell me which data source was used to produce the results (EIL-2008 is the name of the dataset) and important information about the overall analysis (e.g., that robust standard errors were used, what the star levels mean, and which explanatory variables were included in these regression models). Even if I only read the abstract for Iriondo's study before turning to this table, I would have a general idea of what the results indicated.

Iriondo also provides an interpretation of the results in this table in the text of his article: "Table 7 [10.1 in this chapter] shows results obtained from the UCM graduate sample. [...] The results indicate that Erasmus mobility had no significant short-term effect on [...] employment [...]. In contrast, 6 years later, Erasmus graduates were observed to have a 10.6 per cent higher probability of having a job [...]" (pp. 935–937). This text serves to direct the reader's attention to the information in the table that directly corresponds to Iriondo's research question about the impact of Erasmus participation on employment, found in the first line of the table. While the other information in the table is useful and necessary for communicating the study's full findings, readers need to know where to find the most important pieces of information.

10.1.5 Limitations

After the results section is one of several places where a section about the study's limitations can be most appropriate. In practice, where you place this section can depend on a number of factors, including preference of the publication venue where you intended to send your manuscript, where it makes the most sense in the flow of your paper (e.g., if all of your limitations relate to your dataset, this section might belong best in your Methods section), and your own personal preference. Some researchers prefer to include this section at the very end of a manuscript so as not to disrupt the flow of the article itself. Regardless of where your Limitations section ends up, all research has limitations, and it is important to acknowledge them somewhere in your paper. Common limitations in education research include lack of information for key variables related to the study (this is an especially common problem when using secondary data), a small number of study participants in particular groups that are key to the purpose of the study (e.g., a low number of students from specific demographic groups who participated in study abroad), specificity of the data to a particular kind of higher education institution or even a single institution (that is, lack of generalizability outside the specific study context), specificity to a particular geographic area (again, lack of generalizability), and presence of missing data for important variables. Of course, each study is different and comes with its own unique limitations.

10.1.6 Discussion and Conclusion

The Discussion section of an academic paper serves to place the study's findings in the context of prior theory and research and discusses what these findings mean. I generally recommend that Discussion sections contain roughly three main parts. The first section summarizes the key findings of the study and describes how they relate to the research questions or hypotheses that were outlined at the beginning of the paper. This section might also bring these key findings into conversation with findings from previous literature as well as the theoretical framework that informed the study, but this conversation often makes more sense as its own, second section. The third section of your Discussion summarizes implications of the research. Implications can fall into three main categories—implications for future research, implications for theory, and implications for policy and practice. Your study may speak to one of these categories of implications more than another, but it is useful to consider all three when brainstorming the content of this section. Finally, your Conclusion section should review and reiterate why your study and its findings are important, perhaps highlighting one or two key takeaway messages that you want your readers to remember.

10.2 Writing for External Audiences

While the previous section outlines how to write about quantitative research for a primarily academic audience, readers of this book may find that they also need to write for other kinds of audiences, whether instead of or in addition to academic writing. These external audiences likely fall into two main groups: policymakers and practitioners. The format for conveying research results to these audiences can vary considerably, depending on your specific audience and the venue for your publication. To this end, the best way to know what content expectations are for this kind of written product is to look at other publications from the same organization. Some organizations, such as NAFSA: Association for International Educators or the Association for International Education Administrators, often have their own templates or formatting guidelines for research briefs and other research-related publications. Here, I provide a list of recommendations to consider when writing for these external audiences[5]:

1. Articulate why your research matters and what it means—just stating a fact or finding using data does not necessarily tell a story that is interesting or useful for policymakers and practitioners.
2. Provide clear and concrete recommendations—what do you want policymakers or practitioners to do?
3. Think about the story you are trying to tell and how you can make that story real for your audience. Can you share an example or a narrative that helps your readers contextualize the issue at hand?
4. Keep your writing simple and direct with no assumptions of prior knowledge or experience. Write in a way that your grandparents would understand.
5. Identify and keep in mind the needs of your readers and adapt the content and language of your writing to them.
6. Keep it brief and get to your point quickly—assume that your audience will not read more than a page, then assume they will not read more than a paragraph, and then assume that they will not read more than a few bullet points or maybe even more than the title. In other words, make these smaller pieces of text count.
7. Use bullet points.
8. Focus on relevant content and use language that your audience is used to reading. This means that elements of academic writing, such as technical language related to theory or research methods or equations, have very little place (if any) in writing for policymakers or practitioners. If you feel like you need to include details about your analysis, you might consider including these in an appendix rather than your main text.
9. Look at websites and publications of your target venue and find out what sorts of communication your audience is used to receiving.
10. Include pictures and graphs to covey your main points. Assume that your readers may only look at the pictures and graphs and not read the associated written content.
11. Do some background research on existing policy and practice in your topic area so that your recommendations are in-line with what your readers are already thinking about. Consider how your recommendations could conflict or combine with current recommendations to create unintended consequences.

Because the content and structure of research-focused writing for non-researcher audiences varies widely, I am unable to provide a general template for these sorts of publications. I hope that this list of recommendations provides some guidance for beginning to think about what these written products are like.

---

[5] I am especially grateful to my Twitter followers, who helped me crowdsource many of these recommendations.

Practice Problems Answer Key

Chapter 2:

1. The following table contains information about the number of hours per week a group of ten students spent studying for their foreign language class. Use this table to calculate by hand the following pieces of information:
    a. Mode $= 0$
    b. Median $= \frac{10+12}{2} = 11$
    c. Mean $= \frac{10+15+20+0+3+25+23+9+12+0}{10} = 11.7$

Student	Hours studying
1	10
2	15
3	20
4	0
5	3
6	25
7	23
8	9
9	12
10	0

Note Data in this table are invented for illustrative purposes

d. Is this distribution skewed? If so, in which direction? How do you know?

This distribution is right skewed because the median is lower than the mean.

© The Editor(s) (if applicable) and The Author(s), under exclusive license to Springer Nature Switzerland AG 2022
M. Whatley, *Introduction to Quantitative Analysis for International Educators*, Springer Texts in Education, https://doi.org/10.1007/978-3-030-93831-4

2. Sample Dataset #1 contains information about the number of students who studied abroad through U.S. flagship institutions in the 2015–16 academic year (studyabroad). Using your statistical software program of choice, find the mode, median, and mean of this variable. Is this distribution skewed? If so, in what direction? How do you know?

The median of this distribution is 1112.5 and the mean is 1276.68. There is no mode, as each value for study abroad participation is unique. This distribution is right skewed because the median is lower than the mean.

Chapter 3:

1. You began working with the following table in the previous chapter, which contains information about the number of hours per week a group of ten students spent studying for their foreign language class.

Student	Hours studying
1	10
2	15
3	20
4	0
5	3
6	25
7	23
8	9
9	12
10	0

Note Data in this table are invented for illustrative purposes

You have already calculated measures of central tendency for this dataset by hand. Now, calculate the following pieces of information:

a. Range = 25 – 0 = 25
b. Variance

The table below shows the calculations needed to arrive at the variance for this problem. Remember that the equation for variance is $\frac{\Sigma(X_i - \overline{X})^2}{n-1}$. Note that the mean of this variable ($\overline{X}$) is 11.7, as calculated in Practice Problem 1c.

$X_i - \overline{X}$	$(X_i - \overline{X})^2$
10 – 11.7 = –1.7	2.89
15 – 11.7 = 3.3	10.89
20 – 11.7 = 8.3	68.89

Practice Problems Answer Key

$X_i - \overline{X}$	$(X_i - \overline{X})^2$
$0 - 11.7 = -11.7$	136.89
$3 - 11.7 = -8.7$	75.69
$25 - 11.7 = 13.3$	176.89
$23 - 11.7 = 11.3$	127.69
$9 - 11.7 = -2.7$	7.29
$12 - 11.7 = 0.3$	0.09
$0 - 11.7 = -11.7$	136.89

The sum of the second column in this table $((X_i - \overline{X})^2)$ is 744.1. To arrive at the variance, we divide by $n - 1$, or 9: $\frac{744.1}{9} = 82.68$.

c. Standard Deviation

The standard deviation is the square root of the variance: $\sqrt{82.68} = 9.09$

2. We began working with Sample Dataset #1 in the previous chapter, taking a look at measures of central tendency for the number of students who studied abroad through U.S. flagship institutions in the 2015–16 academic year (studyabroad). Using your statistical software program of choice, now find the range, variance, and standard deviation of this variable. If your software has plotting functions, plot the distribution to visualize this variable.

Range	$3019 - 40 = 2979$
Variance	657,293.7
Standard Deviation	810.7365

Histogram of studyabroad

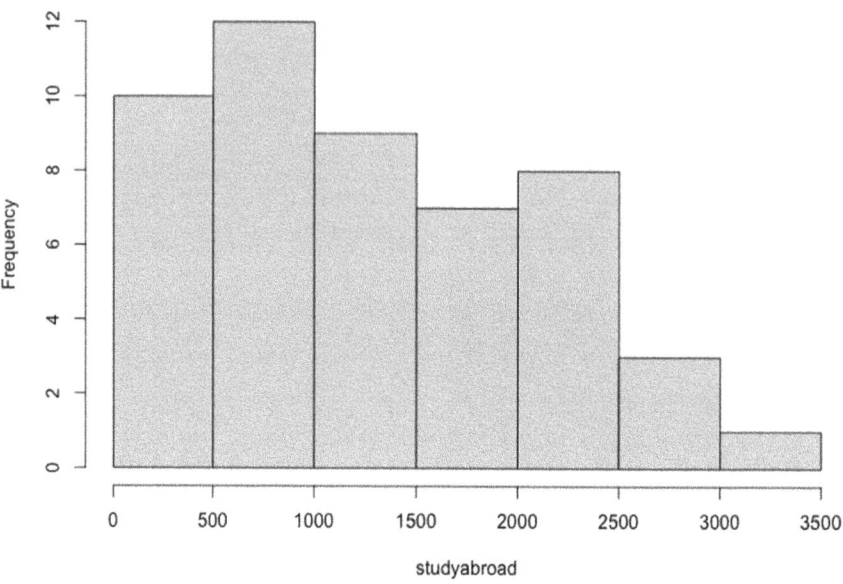

Chapter 4:

1. An administrator makes the claim that international students at my institution have an average GPA of 2.5. To test if this hypothesis is correct, I draw a random sample of 10 students and find that their average GPA is 3.09 with a standard deviation of 0.57.

   a. What are the null and alternative hypotheses (two-tailed) for a comparison between this sample GPA and the population average?

   H_0: The mean GPA of international students is 2.5.
   H_A: The mean GPA of international students is not 2.5.

   b. What is the estimated standard error of the mean for this distribution?

   $$s_{\bar{x}} = \frac{s}{\sqrt{n}} = \frac{0.57}{\sqrt{10}} = 0.18$$

   c. What is the calculated t needed to test the null hypothesis in (a)?[1]

   $$t = \frac{\bar{X} - \mu_0}{s_{\bar{x}}} = \frac{3.09 - 2.5}{0.18} = 3.28$$

---

[1] Note that if you do not round the standard error in part (b) to include in the denominator here, your answer will round to 3.27.

d. What is the critical t value needed to test the null hypothesis in (a) if $\alpha = 0.05$?

2.2622 (Note that this value is found in Appendix A when $df = 9$)

e. Do you reject or fail to reject the null hypothesis?

In this case, we reject the null hypothesis and conclude that the mean GPA of international students is significantly different from the hypothesized population mean GPA of 2.5 ($\alpha = 0.05$).

f. Construct a 95% confidence interval for the sample mean.

$$CI = \overline{X} \pm (t_\alpha)(s_{\overline{x}}) = 3.09 \pm (2.2622)(0.18) = 3.09 \pm 0.4072 = [2.68, 3.49]$$

2. The table below provides information about the number of globalized courses two groups of community college students took during their degree programs. On average, students in arts and sciences degree programs took more globalized courses ($\overline{X}_1 = 2.8$) compared to those in applied degree programs ($\overline{X}_2 = 1.3$).

	Arts and sciences degree programs	Applied degree programs
Average number of globalized courses	2.8	1.3
N	100	100
SE	0.3	0.8

Note Data in this table are invented for illustrative purposes

a. What are the null and alternative hypotheses (two-tailed) for a comparison between the mean number of globalized courses taken by these two groups of students?

H_0: The difference in mean number of globalized courses taken by students in arts and sciences degree programs and the mean number of globalized courses taken by students in applied degree programs is zero.

H_A: The difference in mean number of globalized courses taken by students in arts and sciences degree programs and the mean number of globalized courses taken by students in applied degree programs is not zero.

b. What is the standard error of the difference between the two sample means?

$$s_{\overline{x}1 - \overline{x}2} = \sqrt{s_{\overline{x}1}^2 + s_{\overline{x}2}^2} = \sqrt{0.3^2 + 0.8^2} = \sqrt{0.09 + 0.64} = \sqrt{0.73} = 0.85$$

c. What is the calculated t needed to test the null hypothesis in (a)?

$$t = \frac{\overline{X}_1 - \overline{X}_2}{s_{\overline{x}1 - \overline{x}2}} = \frac{2.8 - 1.3}{0.85} = 1.76$$

d. What is the critical t value needed to test the null hypothesis in (a) if $\alpha = 0.01$?

2.6603(Note that this value is found in Appendix A when df = 60, a conservative estimate)

e. Do you reject or fail to reject the null hypothesis?

In this case, we fail to reject the null hypothesis that the mean number of globalized courses taken by the two groups of students, those in Arts and Sciences degree programs and those in Applied degree programs, is significantly different from zero ($\alpha = 0.01$).

3. Sample Dataset #2 contains information about study abroad participation (studyabroad) at 161 U.S. liberal arts institutions in the 2016–17 academic year. Using the statistical software of your choice, conduct t-tests that examine the following null hypotheses. Write your conclusions out in words.
   a. The mean study abroad participation at liberal arts institutions is equal to 200.

A one-sample t-test suggested that mean study abroad participation at U.S. liberal arts institutions in the 2016–17 academic year ($\overline{X} = 179.7205$, $s_{\overline{x}} = 10.64903$) was not significantly different from a hypothesized mean of 200 ($p > .05$).

b. The mean study abroad participation at liberal arts institutions is equal to 300.

A one-sample t-test suggested that mean study abroad participation at U.S. liberal arts institutions in the 2016–17 academic year ($\overline{X} = 179.7205$, $s_{\overline{x}} = 10.64903$) was significantly different from a hypothesized mean of 300 ($p < .001$).

c. The difference between mean study abroad participation at liberal arts institutions in the highest SAT category (SAT700plus) and the mean study abroad participation at all other liberal arts institutions is zero.

A two-samples t-test suggested that mean study abroad participation at U.S. liberal arts institutions in the 2016–17 academic year in the highest SAT category ($\overline{X} = 261$, $s_{\overline{x}} = 24.96001$) is significantly different from study abroad participation at all other liberal arts institutions in the same academic year ($\overline{X} = 160.3385$, $s_{\overline{x}} = 11.15887$) ($p < 0.001$).

d. The difference between mean study abroad participation at liberal arts institutions with no reported SAT information (SAT_none) and the mean study abroad participation at all other liberal arts institutions is zero.

A two-samples t-test suggested that mean study abroad participation at U.S. liberal arts institutions in the 2016–17 academic year with no reported SAT information ($\overline{X} = 184.8302$, $s_{\overline{x}} = 16.91989$) was not significantly different from study abroad participation at all other liberal arts institutions in the same academic year ($\overline{X} = 177.213$, $s_{\overline{x}} = 13.57699$) ($p > .05$).

Chapter 5:

1. Decide which statistical test (*t*-test, one-way ANOVA, or chi-square) is appropriate to address the following research questions:
   a. Are institutions located in rural areas (compared to non-rural areas) less likely to offer study abroad opportunities? *Chi-square (both variables are categorical)*
   b. Do institutions that focus primarily on STEM graduate training enroll more international students? *T-test (one variable—STEM training—is binary and the other—number of international students enrolled—is continuous)*
   c. What is the relationship between an institution's geographic location (rural, town, suburban, urban) and the number of international research collaborations (measured as number of publications co-authored with an international collaborator) that its faculty produce? *ANOVA (one variable—geographic location—is categorical with more than two categories and the other—number of international research collaborations—is continuous)*
   d. What is the relationship between an institution's size (small, medium, large) and its likelihood of offering on-campus intercultural training opportunities? *Chi-square (both variables are categorical)*
   e. Do institutions that enroll international students have higher graduation rates compared to those that do not? *T-test (one variable—whether an institution enrolls international students—is binary and the other—graduation rate—is continuous)*
2. The data in the table below provide information about the number of international research collaborations of 18 faculty members located at research universities of three different sizes (small, medium, and large) (note that there are six researchers in each institution size category). Conduct a one-way ANOVA to test the null hypothesis that the mean number of international research collaborations of faculty members at institutions in each size category is the same.

Researcher	Small	Medium	Large	
1	4	5	8	
2	2	4	9	
3	1	1	4	
4	0	7	15	
5	0	8	12	
6	3	12	10	Grand mean
Mean	1.7	6.2	9.7	5.83

Note Data in this table are invented for illustrative purposes

a. Calculate the sum of squares between and the mean of squares between.

Sum of squares between (Note that $\overline{X}_T = 5.83$):

Group	Step 1: Subtract the grand mean from the group mean	Step 2: Square the differences	Step 3: Account for sample size (n) of each group
	$(\overline{X} - \overline{X}_T)$	$(\overline{X} - \overline{X}_T)^2$	$n(\overline{X} - \overline{X}_T)^2$
Small	1.7–5.83 = −4.13	$-4.13^2 = 17.0569$	17.0569*6 = 102.3414
Medium	6.2–5.83 = 0.37	$0.37^2 = 0.1369$	0.1369*6 = 0.8214
Large	9.7–5.83 = 3.87	$3.87^2 = 14.9769$	14.9769*6 = 89.8614
	Step 4: Sum squared differences	$\sum n(\overline{X} - \overline{X}_T)^2$	102.3414 + 0.8214 + 89.8614 = 193.0242

Mean of squares between:

$$MS_b = \frac{193.0242}{3-1} = 96.5121$$

b. Calculate the sum of squares within and the mean of squares within

	Small $(\overline{X} = 1.7)$		Medium $(\overline{X} = 6.2)$		Large $(\overline{X} = 9.7)$	
	Step 1	Step 2	Step 1	Step 2	Step 1	Step 2
	$X - \overline{X}$	$(X - \overline{X})^2$	$X - \overline{X}$	$(X - \overline{X})^2$	$X - \overline{X}$	$(X - \overline{X})^2$
1	4–1.7 = 2.3	5.29	5–6.2 = -1.2	1.44	8–9.7 = -1.7	2.89
2	2–1.7 = 0.3	0.09	4–6.2 = -2.2	4.84	9–9.7 = -0.7	0.49
3	1–1.7 = -0.7	0.49	1–6.2 = -5.2	27.04	4–9.7 = -5.7	32.49
4	0–1.7 = -1.7	2.89	7–6.2 = 0.8	0.64	15–9.7 = 5.3	28.09
5	0–1.7 = -1.7	2.89	8–6.2 = 1.8	3.24	12–9.7 = 2.3	5.29
6	3–1.7 = 1.3	1.69	12–6.2 = 5.8	33.64	10–9.7 = 0.3	0.09
Step 3: Sum the squared differences	$\sum(X - \overline{X})^2 =$	13.34	$\sum(X - \overline{X})^2 =$	70.84	$\sum(X - \overline{X})^2 =$	69.34

$$SS_w = 13.34 + 70.84 + 69.34 = 153.52$$

$$MS_w = \frac{153.52}{18 - 3} = 10.24$$

Practice Problems Answer Key

c. Find the calculated F-statistic.

$$F = \frac{MS_b}{MS_w} = \frac{96.5121}{10.24} = 9.43$$

d. Find the critical F-statistic ($\alpha = .05$).

3.6823(Note that this value is found in Appendix B when df in the numerator = 2 and df in the denominator = 15)

e. Do you reject or fail to reject the null hypothesis?

In this case, we reject the null hypothesis as we have evidence that the number of international research collaborations of faculty members is significantly related to the size of the university where they are employed ($p < 0.05$).

3. Using Sample Dataset #1 and the statistical software program of your choice, explore the null hypothesis that mean international student enrollment (intlstudents) at U.S. flagship institutions is the same for all institutional locales (rural, town, suburban, urban) ($\alpha = 0.05$).

A one-way ANOVA suggested that we do not have evidence to reject the null hypothesis–international student enrollment at U.S. flagship institutions appears to be unrelated to an institution's locale (rural, town, suburban, urban) ($F(3,46) = 2.40$, $p > 0.05$).

4. The data in the table below provide information about the distribution of international student enrollment by region of origin (defined as the continent that a student is from) and the size of the institution of higher education where the student enrolled. Conduct a chi-square test to examine the null hypothesis that an international student's continent of origin is not related to the size of the institution where they enroll.

	Small	Medium	Large
Africa	356	746	987
Americas	1356	236	512
Asia	128	153	998
Australia/Oceania	345	321	413
Europe	98	352	732

Note Data in this table are invented for illustrative purposes

a. Find the row and column sums for this table.

	Small	Medium	Large	Sum
Africa	356	746	987	2089
Americas	1356	236	512	2104
Asia	128	153	998	1279
Australia/Oceania	345	321	413	1079
Europe	98	352	732	1182
Sum	2283	1808	3642	**7733**

b. Calculate the expected frequencies of international student enrollment by continent of origin and size of institution.

$$E = \frac{(row\ sum) * (column\ sum)}{N}$$

	Small	Medium	Large	Sum
Africa	$\frac{2089*2283}{7733}$	$\frac{2089*1808}{7733}$	$\frac{2089*3642}{7733}$	2089
Americas	$\frac{2104*2283}{7733}$	$\frac{2104*1808}{7733}$	$\frac{2104*3642}{7733}$	2104
Asia	$\frac{1279*2283}{7733}$	$\frac{1279*1808}{7733}$	$\frac{1279*3642}{7733}$	1279
Australia/Oceania	$\frac{1079*2283}{7733}$	$\frac{1079*1808}{7733}$	$\frac{1079*3642}{7733}$	1079
Europe	$\frac{1182*2283}{7733}$	$\frac{1182*1808}{7733}$	$\frac{1182*3642}{7733}$	1182
Sum	2283	1808	3642	7733

	Small	Medium	Large	Sum
Africa	616.73	488.41	983.85	2089
Americas	621.16	491.92	990.92	2104
Asia	377.60	299.03	602.37	1279
Australia/Oceania	318.55	252.27	508.18	1079
Europe	348.96	276.36	556.68	1182
Sum	2283	1808	3642	7733

c. Find the calculated chi-square value.

$$\chi^2 = \sum \left(\frac{(O-E)^2}{E} \right)$$

Practice Problems Answer Key

	Small		Medium		Large	
	Observed	Expected	Observed	Expected	Observed	Expected
Africa	356	616.73	746	488.41	987	983.85
Americas	1356	621.16	236	491.92	512	990.92
Asia	128	377.60	153	299.03	998	602.37
Australia/Oceania	345	318.55	321	252.27	413	508.18
Europe	98	348.96	352	276.36	732	556.68

	Small		
	$O - E$	$(O - E)^2$	$\frac{(O-E)^2}{E}$
Africa	−260.73	67,980.13	110.23
Americas	734.84	539,989.83	869.32
Asia	−249.60	62,300.16	164.99
Australia/Oceania	26.45	699.60	2.20
Europe	−250.96	62,980.92	180.48

	Medium		
	$O - E$	$(O - E)^2$	$\frac{(O-E)^2}{E}$
Africa	257.59	66,352.61	135.85
Americas	−255.92	65,495.05	133.14
Asia	−146.03	21,324.76	71.31
Australia/Oceania	68.73	4723.81	18.72
Europe	75.64	5721.41	20.70

	Large		
	$O - E$	$(O - E)^2$	$\frac{(O-E)^2}{E}$
Africa	3.15	9.92	0.01
Americas	−478.92	229,364.37	231.47
Asia	395.63	156,523.10	259.85
Australia/Oceania	−95.18	9059.23	17.83
Europe	175.32	30,737.10	55.21

$$\chi^2 = \sum \left(\frac{(O - E)^2}{E} \right) = 110.23 + 869.32 + 164.99 + 2.20 + 180.48$$
$$+ 135.85 + 133.14 + 71.31 + 18.72 + 20.70 + 0.01 + 231.47$$
$$+ 259.85 + 17.83 + 55.21 = 2271.32$$

d. Find the critical chi-square value when $\alpha = 0.05$.

15.5073 (Note that this value is found in Appendix C when $df = (5-1)*(3-1) = 8$)

e. Do you reject or fail to reject the null hypothesis?

In this case, we reject the null hypothesis and conclude that region of origin of enrolled international students differs significantly according to the size of the institution (small, medium, or large) ($p < 0.05$).

5. Using Sample Dataset #2 and the statistical software program of your choice, explore the null hypothesis that whether a liberal arts college offers study abroad (studyabroad_offered) is not related to its geographic locale (rural, town, suburban, urban) ($\alpha = 0.05$).

A chi-square test of independence suggests that whether a liberal arts college offers study abroad is not significantly related to its geographic locale (rural, town, suburban, urban) ($\chi^2(3) = 7.36, p > 0.05$).

Chapter 6:

1. The following table provides the raw data used to produce Fig. 6.1, which visually displayed the relationship between internationally collaborative publications and grant funding.

Institution	Internationally collaborative publications	Grant funding amount (in millions)
A	27	0.5
B	236	7
C	104	1
D	54	2
E	128	4
F	114	3.5
G	21	7
Mean	97.71	3.57

Use this table to calculate the following three pieces of information about this correlation by hand:

a. The correlation coefficient

First, calculate the covariance between the two variables represented in the table (in this case, internationally collaborative publications is x and grant funding amount is y):

$$S_{xy} = \frac{\sum(X - \overline{X})(Y - \overline{Y})}{n - 1}$$

Practice Problems Answer Key

	Intl publications	$X - \bar{X}$	Grant funding	$Y - \bar{Y}$	$(X - \bar{X})(Y - \bar{Y})$
A	27	27–97.71 = -70.71	0.5	0.5–3.57 = -3.07	217.08
B	236	236–97.71 = 138.29	7	7–3.57 = 3.43	474.33
C	104	104–97.71 = 6.29	1	1–3.57 = -2.57	-16.17
D	54	54–97.71 = -43.71	2	2–3.57 = -1.57	68.62
E	128	128–97.71 = 30.29	4	4–3.57 = 0.43	13.02
F	114	114–97.71 = 16.29	3.5	3.5–3.57 = -0.07	-1.14
G	21	21–97.71 = -76.71	7	7–3.57 = 3.43	-263.12
					Sum = 492.64

$$s_{xy} = \frac{492.64}{7 - 1} = 82.11$$

Remember that the denominator in the formula for the correlation coefficient uses the standard deviations for variables x and y separately, calculated using the formula $s = \sqrt{\frac{\sum(X_i - \bar{X})^2}{n-1}}$. *This information is easily calculated with the information already in the previous table:*

	Intl publications	$(X - \bar{X})$	$(X - \bar{X})^2$
A	27	27–97.71 = -70.71	4999.90
B	236	236–97.71 = 138.29	19,124.12
C	104	104–97.71 = 6.29	39.56
D	54	54–97.71 = -43.71	1910.56
E	128	128–97.71 = 30.29	917.48
F	114	114–97.71 = 16.29	265.36
G	21	21–97.71 = -76.71	5884.42
			Sum = 33,141.43

$$s_x = \sqrt{\frac{33141.43}{7 - 1}} = 74.32$$

	Grant funding	$(Y - \bar{Y})$	$(Y - \bar{Y})^2$
A	0.5	0.5–3.57 = -3.07	9.42
B	7	7–3.57 = 3.43	11.76
C	1	1–3.57 = -2.57	6.60
D	2	2–3.57 = -1.57	2.46
E	4	4–3.57 = 0.43	0.18
F	3.5	3.5–3.57 = -0.07	0.00
G	7	7–3.57 = 3.43	11.76
			Sum = 42.18

$$s_x = \sqrt{\frac{42.18}{7-1}} = 2.65$$

Once you have calculated these standard deviations, use these values along with the covariance to calculate the correlation coefficient:

$$r_{xy} = \frac{s_{xy}}{(s_x)(s_y)} = \frac{82.11}{(74.32)(2.65)} = \frac{82.11}{196.95} = 0.42$$

b. The coefficient of determination

The coefficient of determination is the square of the correlation coefficient: $0.42^2 = 0.18$

c. Whether the correlation coefficient is significantly different from zero

To evaluate whether the correlation coefficient (0.42) is significantly different from zero, you first need to calculate the standard error of the sample correlation coefficient, as it is the denominator in the calculation of t:

$$s_r = \sqrt{\frac{(1-r^2)}{(n-2)}} = \sqrt{\frac{(1-0.42^2)}{(7-2)}} = \sqrt{\frac{(1-0.1764)}{5}} = \sqrt{\frac{0.8236}{5}} = \sqrt{0.16472} = 0.41$$

Using this value in the denominator, you can then calculate the t-value you will need for a hypothesis test:

$$t = \frac{r - \rho}{s_r} = \frac{0.42 - 0}{0.41} = 1.02$$

The critical t-value in this case, found in Appendix A, is associated with $df = 5$ ($N-2$). If we assume alpha $= 0.05$, then the critical t-value is 2.5706. Because this value is greater than our calculated t-value, we fail to reject the null hypothesis.

2. Sample Dataset #1 contains information about international student enrollment numbers (intlstudents) and total enrollment (totalenroll). Figure 6.5 visualizes these two variables.

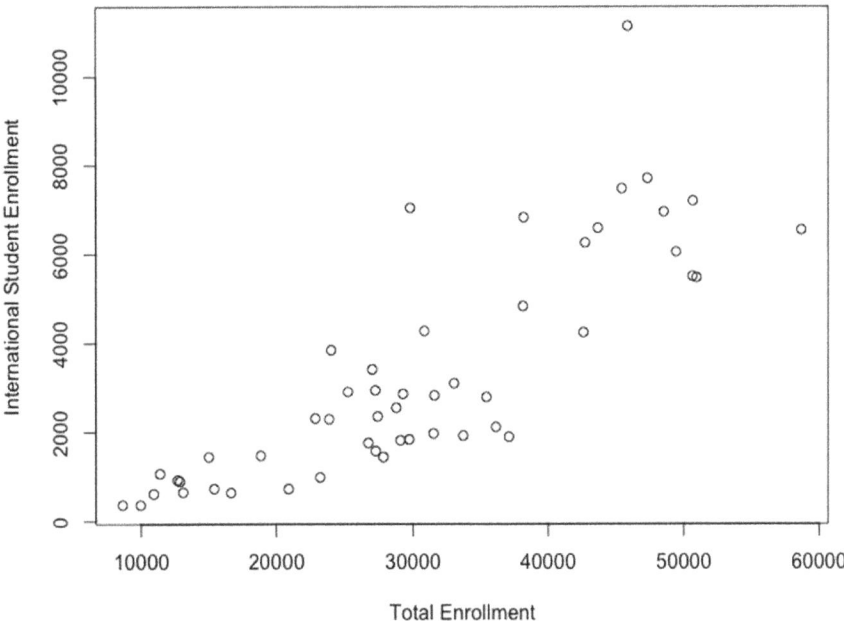

Fig. 6.5 Correlation Between International Student Enrollment and Total Enrollment at U.S. Flagship Institutions from the 2015–16 Academic Year. *Data source*: Sample Dataset #1-US National Center for Education Statistics, Integrated Postsecondary Education Data System)

Relying on Fig. 6.5, do you expect the correlation between these two variables to be positive or negative? Why?

You would expect the correlation between these two variables to be positive, as the pattern in the figure suggests that, on average, when enrollment increases, so does international student enrollment.

3. Using this dataset and your statistical software program of choice, calculate the following pieces of information about the relationship between international student enrollment numbers and a flagship institution's total enrollment:
a. The correlation coefficient

$$r_{xy} = 0.8241$$

b. The coefficient of determination

$$r_{xy}^2 = 0.6791$$

c. Whether the correlation coefficient is significantly different from zero

The correlation coefficient in this example ($r_{xy} = 0.8241$) is significantly different from zero ($p < 0.001$).

Chapter 7:

1. Sample Dataset #2 contains information from 161 U.S. Liberal Arts institutions from the 2016–17 academic year. Here, we focus on the number of students who studied abroad from these institutions (represented by the variable studyabroad). The table below summarizes an OLS regression model that takes the following form:

$$TotalSA_i = a + b_1 TotalEnroll1000_i + b_2 PctFemale_i + b_3 AcceptRate_i + e_i$$

This regression uses a liberal arts institution's total enrollment (measured in 1000s) (totalenroll), percentage of students identifying as female (pctfemale), and acceptance rate (acceptrate) to predict total study abroad participation.

Variable	Coefficient
(Intercept)	259.38***
	(48.20)
Total Enrollment (in 1000s)	36.36***
	(7.45)
Percent female	−0.60
	(0.65)
Acceptance rate	−2.12***
	(0.44)
N observations	161
R^2	0.23
Adjusted R^2	0.22

Note' Standard errors in parentheses. ***$p < 0.001$. Data source: Sample Dataset #2-US National Center for Education Statistics, Integrated Postsecondary Education Data System

a. Interpret the coefficients corresponding to total enrollment, percent female, and acceptance rate in this regression model. Which ones are statistically significant?

A 1000 increase in total enrollment is associated with an approximate increase of 36 students, on average, participating in study abroad ($p < 0.001$). While a one-point increase in the percentage of students identifying as female is associated with

Practice Problems Answer Key

approximately one fewer student studying abroad, on average, this predictor variable is not significant at a standard level. Regarding acceptance rate, a one-point increase in acceptance rate is associated with approximately two fewer students, on average, participating in study abroad ($p < 0.001$).

b. What percentage of the variance in study abroad participation does this regression model explain?

This regression model explains approximately 22% of the variance in study abroad participation, as suggested by the adjusted R^2 value.

c. Write out the equation that corresponds to this regression model.

$$\widehat{TotalSA_i} = 259.38 + 36.36 * TotalEnroll1000_i - 0.60 * PctFemale_i - 2.12 * AcceptRate_i$$

What is the predicted number of students who study abroad at liberal arts institutions with the following characteristics:

i. Total Enrollment = 7,000, Percent female = 67%,[2] Acceptance rate = 20%

431.30

ii. Total Enrollment = 300, Percent female = 55%, Acceptance rate = 10%

216.09

iii. Total Enrollment = 750, Percent female = 98%, Acceptance rate = 5%

217.25

2. Now use your statistical software program of choice to load Sample Dataset #2 and run this regression model yourself (note that you will have to create the enrollment in 1000s variable yourself). Double-check your work to ensure that you arrived at the same numbers as in the table above. In a second step, add two additional predictor variables to the model: the percentage of the student body that is comprised of graduate students (pctgrad) and the percentage of the student body receiving Pell grants (pctpell). The equation version of this model is as follows:

$$TotalSA_i = a + b_1 TotalEnroll1000_i + b_2 PctFemale_i + b_3 AcceptRate_i$$

---

[2] Note that numbers for percent female and acceptance rate should be taken as percentages (67%) rather than proportions (.67) for the purposes of this practice problem.

$$+ b_4 PctGrad_i + b_5 PctPell_i + e_i$$

a. Interpret the coefficients corresponding to total enrollment, percent female, acceptance rate, percent graduate student, and percent Pell recipient in this regression model. Which ones are statistically significant?

The table below provides the results of this second regression model. This table suggests that a 1000-student increase in total enrollment is associated with an increase of approximately 38 students studying abroad, on average. On the other hand, a one-point increase in the percentage of students receiving Pell is associated with approximately 6 fewer students participating in study abroad, on average. Both these variables are statistically significant at a standard level ($p < 0.001$). The other variables, percent female, acceptance rate, and percent graduate students, are not significant predictors of study abroad participation.

Variable	Coefficient
(Intercept)	314.24***
	(43.65)
Total Enrollment (in 1000s)	38.13***
	(6.54)
Percent female	−0.23
	(0.57)
Acceptance rate	−0.75
	(0.42)
Percent graduate students	−0.97
	(0.91)
Percent Pell	−6.05***
	(0.83)
N observations	161
R^2	0.44
Adjusted R^2	0.42

Note Standard errors in parentheses. ***$p < 0.001$. Data source: Sample Dataset #2-US National Center for Education Statistics, Integrated Postsecondary Education Data System

a. Are your results for total enrollment, percent female, and acceptance rate different from the first model you ran? Why?

The results for total enrollment, percent female, and acceptance rate are somewhat different from the first model because we have added additional predictors to the model.

b. Does this second model explain more or less of the variance in study abroad participation compared to the first model? Why? How does the penalized fit from the adjusted R^2. compare?

This model explains more of the variance in study abroad participation compared to the first model because we have added more predictors. While the adjusted R^2 for this model is still higher than the first model, it seems to be penalized a bit more (rather than one tenth of a point lower, as in the first model, this adjusted R^2 is two tenths of a point lower), likely because there are more predictors in this model.

c. If you were the Director of Study Abroad at a liberal arts institution, how could you use the results of this regression model in your work?

Although the results of this model are certainly preliminary and should be interpreted with caution, of note is that the percentage of students receiving Pell is negatively related to study abroad participation. That is, liberal arts institutions that serve a higher percentage of students from low-income backgrounds report fewer students studying abroad. This result should encourage international educators to think more deeply about how they can better serve this student population.

Chapter 8:

1. Sample Dataset #2 contains information about 161 Liberal Arts institutions in the 2016–17 academic year. In this first practice activity, we focus again on the number of students who studied abroad from these institutions (studyabroad). The table below summarizes an OLS regression model that builds on the one introduced in practice activity 1 in the previous chapter. This regression model adds three variables to this equation: one that allows the acceptance rate variable to depict a quadratic functional form ($AcceptRate_i^2$), a dummy variable indicating whether the institution is public ($Public_i$) and a final variable that represents the interaction between $Public_i$ and $TotalEnroll1000_i$ ($Public * TotalEnroll1000_i$).

$$TotalSA_i = a + b_1 TotalEnroll1000_i + b_2 PctFemale_i + b_3 AcceptRate_i$$
$$+ b_4 AcceptRate_i^2 + b_5 Public_i + b_5 Public * TotalEnroll1000_i + e_i$$

Variable	Coefficient
(Intercept)	137.85*
	(55.89)
Total Enrollment (in 1000s)	116.77***
	(11.55)

Variable	Coefficient
Percent female	−0.93 +
	(0.54)
Acceptance rate	−3.21 +
	(1.70)
Acceptance rate squared	0.02
	(0.02)
Public	87.95 +
	(50.09)
Public*Total Enrollment (in 1000s)	−98.83***
	(14.79)
N observations	161
R^2	0.48
Adjusted R^2	0.46

Note Standard errors in parentheses. + $p < 0.10$, *$p < 0.05$, ***$p < 0.001$. Data source: Sample Dataset #2-US National Center for Education Statistics, Integrated Postsecondary Education Data System

a. Is there evidence that the functional form depicting the relationship between acceptance rate and study abroad participation is quadratic? Why or why not?

There is little evidence to suggest that the relationship between acceptance rate and study abroad participation is quadratic. The squared value for acceptance rate in this regression model is not statistically significant.

i. What diagnostic test might you conduct to determine whether this functional form is appropriate for your data?

To determine if this functional form is appropriate for the data, you can plot the residuals for two regression models: one with only the regular acceptance rate predictor and another that adds a squared term. The model with residuals that cluster more closely around zero is probably the best fit for your data. The figure below shows the residuals for these two models in the context of this practice problem. As expected, the residuals for the two models are similar in appearance.

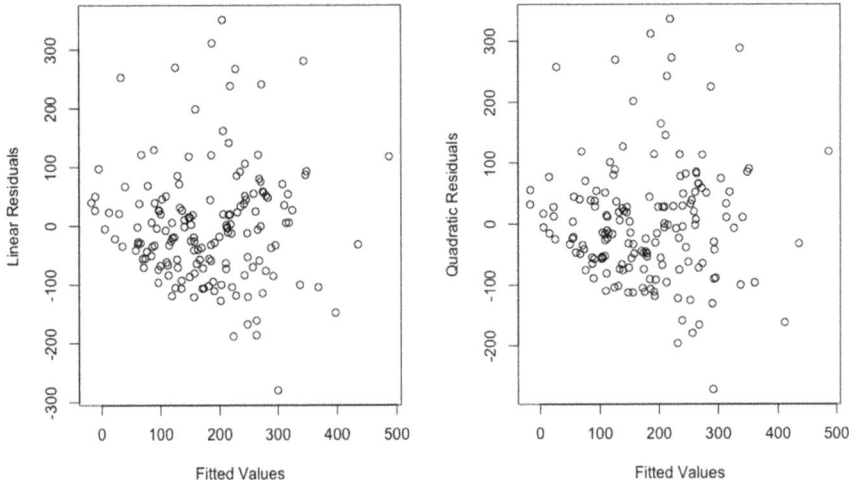

b. How do you interpret the coefficient corresponding to $Public_i$?

The coefficient on the Public variable ($b = 87.95$) suggests a positive relationship between public control and study abroad participation (when enrollment is 0, given that this regression model also has an interaction term between Public and enrollment). If we consider $p < 0.10$ to be statistically significant, this result suggests that, compared to private liberal arts institutions, public liberal arts institutions report, on average, around 88 more students participating in study abroad.

c. How do you interpret the coefficient corresponding to $Public*TotalEnroll1000_i$?

This variable is significant at a standard level ($p < 0.001$), suggesting that there is an important interaction between public control and total enrollment when it comes to study abroad participation.

i. What might you do to explore further the nature of this interaction term?

To further interpret this interaction term, we can plot the predicted values of study abroad participation for public and private institutions as follows:

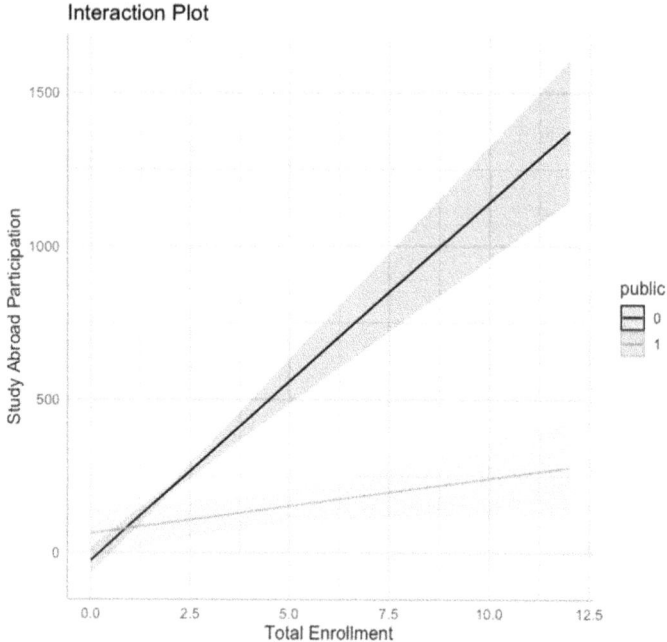

Interaction Plot

This plot suggests that while the relationship between total enrollment and study abroad participation is positive for both public and private institutions, this relationship is significantly stronger for private institutions (public = 0).

d. What percentage of the variance in study abroad participation does this regression model explain?

This regression model explains approximately 46% of the variance in study abroad participation, as suggested by the adjusted R^2 value.

2. Now use your statistical software program of choice to load Sample Dataset #2 and run this regression model yourself. Double-check your work to ensure that you arrived at the same numbers as in the table above. In a second step, consider a quadratic functional form to depict the relationship between total enrollment and study abroad participation using a simpler version of this regression model. Estimate this model using the following equation:

$$TotalSA_i = a + b_1 TotalEnroll1000_i + b_2 TotalEnroll1000_i^2 + b_3 PctFemale_i + b_4 AcceptRate_i + b_5 Public_i + e_i$$

a. Interpret the coefficient corresponding to total enrollment and total enrollment squared. Is there evidence that the relationship between total enrollment and study abroad participation is quadratic? How do you know?

Practice Problems Answer Key

The table below corresponds to this regression model. Both total enrollment predictors (both the linear predictor and the quadratic) are significant, suggesting a quadratic relationship between this variable and study abroad participation. The coefficient on the main term for total enrollment is positive and the coefficient on the squared enrollment term is negative, indicating an inverse-u-shaped relationship between total enrollment and study abroad participation. That is, the relationship between total enrollment and study abroad participation is a positive one, but only until a certain point. After this point, this relationship levels off and may even become negative.

Variable	Coefficient
(Intercept)	100.78*
	(46.71)
Total enrollment (in 1000s)	138.49***
	(16.30)
Total enrollment squared (in 1000s)	−10.13***
	(1.79)
Percent female	−0.92
	(0.56)
Acceptance rate	−1.34**
	(0.40)
Public	−154.50***
	(33.03)
N observations	161
R^2	0.44
Adjusted R^2	0.42

Note Standard errors in parentheses. *$p < 0.05$, **$p < 0.01$, ***$p < 0.001$. Data: Sample Dataset #2-US National Center for Education Statistics, Integrated Postsecondary Education Data System

i. What diagnostic test might you conduct to determine whether this functional form is appropriate for your data?

To determine if this functional form is appropriate for the data, you can plot the residuals for two regression models: one with only the regular total enrollment predictor and another that adds a squared term. The model with residuals that cluster more closely around zero is probably the best fit for your data. The figure below shows the residuals for these two models in the context of this practice problem. In this case, the difference between the two plots is minimal.

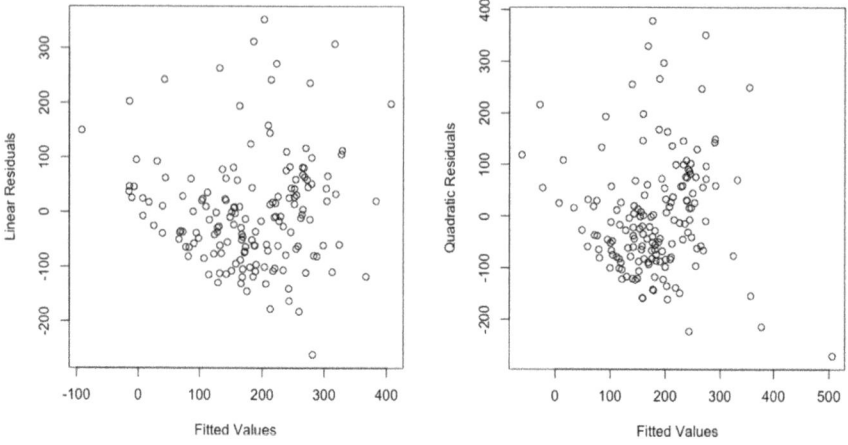

b. What lessons might international educators derive from the regression models in practice problems 1 and 2?

These regression models consistently suggest a positive relationship between total enrollment and study abroad participation. This finding is not especially surprising—institutions that enroll more students have a larger student population to recruit from for study abroad participation. For international educators, this finding provides an important lesson for how we discuss which institutions excel at study abroad participation—conversations that privilege number of study abroad participants inherently put larger institutions at an advantage, perhaps undeservedly.

3. Sample Dataset #3 contains the community college data used to derive the logistic regression examples used in this chapter. Use this dataset to build on the regression model in Table 8.7, which takes offering study abroad as the outcome of interest (studyabroad_offered), so that it accounts for three additional predictor variables (Notice that this regression model will contain 915 observations due to missing data on tuition charges and locale from some community colleges. Be wary of missing data when you create variables for this exercise). Namely, estimate a logistic regression model that takes the following form:

$$\ln\left(\frac{P_i}{1 - P_i}\right) = a + b_1 TotalEnroll1000_i + b_2 Rural_i$$
$$+ b_3 PctOver25_i + b_4 Tuition1000_i$$

In this equation, $Rural_i$ is a binary indicator of whether the institution is located in a rural area (this variable can be derived from locale—see the Data Dictionaries at the beginning of this book for variable definitions), $PctOver25_i$ is the percentage of the student population aged 25 or older (pctover25), and $Tuition1000_i$ is the amount charged in tuition at each college in US$1000 units (this variable can be derived from tuition).

The logistic regression model with unconverted log odds coefficients is displayed in the following table. While this model is the standard output for a logistic regression that most statistical software programs provide, the coefficients are not readily interpreted.

Variable	Coefficient
(Intercept)	−2.30***
	(0.51)
Total Enrollment (in 1000s)	0.09***
	(0.39)
Rural	−0.84*
	(0.39)
Percent over 25	0.00
	(0.01)
Tuition (in 1000s)	−0.15*
	(0.06)
N observations	915
McFadden's Pseudo R^2	0.15

Note Standard errors in parentheses. *$p < 0.05$, ***$p < 0.001$. Data source: Sample Dataset #2-US National Center for Education Statistics, Integrated Postsecondary Education Data System

a. Convert the log odds coefficients from this regression model into odds ratios. How do you interpret the odds ratios corresponding to total enrollment, rural, percent 25+ and tuition charges?

The table below displays the same regression model, but with the log odds coefficients converted into odds ratios. This regression model suggests that a 1000 increase in total enrollment is associated with an approximate 10% increase in the odds of a community college offering study abroad. Conversely, if a college is located in a rural area, we expect a 57% decrease in the odds that study abroad is offered (1−0.43 = 0.57). The percentage of students over 25 has no discernible effect. Finally, a $1000 increase in tuition is related to a 14% decrease in the odds that study abroad is offered. All these findings are significant at $p < 0.05$ or less.

Variable	Odds ratio
(Intercept)	0.10***
	(0.05)
Total enrollment (in 1000s)	1.10***
	(0.02)
Rural	0.43*

Variable	Odds ratio
	(0.17)
Percent over 25	1.00
	(0.01)
Tuition (in 1000s)	0.86*
	(0.05)
N observations	915
McFadden's Pseudo R^2	0.15

Note Standard errors in parentheses. *$p < 0.05$, ***$p < 0.001$. Data source: Sample Dataset #2-US National Center for Education Statistics, Integrated Postsecondary Education Data System

b. Convert the log odds coefficients from this regression model into average marginal effects. How do you interpret the probability changes corresponding to total enrollment, rural, percent 25+ and tuition charges?

The table below displays the same regression model, but with the log odds coefficients converted into average marginal effects. This regression model suggests that a 1000 increase from the mean total enrollment is associated with an increase of about one percentage point in the probability that a community college would offer study abroad. Conversely, if a college is located in a rural area, we expect a decrease in the probability of offering study abroad of around 7 percentage points. Finally, a $1000 increase from the mean tuition is related to an approximate decrease of one percentage point in the probability of a college offering study abroad. All these findings are significant at $p < 0.05$ or less.

Variable	Average marginal effects
Total enrollment (in 1000s)	0.01***
	(0.00)
Rural	−0.07*
	(0.03)
Percent over 25	0.00
	(0.00)
Tuition (in 1000s)	−0.01*
	(0.01)
N observations	915
McFadden's Pseudo R^2	0.15

Note Standard errors in parentheses. *$p < 0.05$, ***$p < 0.001$. Data source: Sample Dataset #2-US National Center for Education Statistics, Integrated Postsecondary Education Data System

c. How well does this regression model fit the data? How do you know?

While the Pseudo R^2 in a logistic regression does not suggest percentage of explained variation in the outcome variable like R^2 in an ordinary least squares regression, this statistic ranges in value from 0 to 1, with values closer to 1 indicating better model fit. Since the Pseudo R^2 in this regression model is closer to 0, it may be that our variables do not account for a number of factors that relate to whether a community college offers study abroad. However, given the complexities of offering study abroad programming, we may not expect an especially high Pseudo R^2 value in any model predicting the likelihood of a community college offering study abroad.

d. What might international educators at community colleges learn from this regression model?

This regression model suggests that community colleges with larger enrollments are more likely to offer study abroad while those in rural areas and those with higher tuition charges are less likely to offer study abroad. International educators at larger community colleges that do not currently offer study abroad might take these results to mean that establishing such programs is a viable option. International educators employed at rural community colleges who want to establish study abroad might look to other rural colleges as partners for the provision of study abroad, perhaps because their colleges are smaller and cannot maintain robust study abroad programming on their own. Finally, international educators at colleges with higher tuition charges might seek out sources of financial aid to offset these charges for students who want to participate in study abroad. Perhaps these higher tuition charges deter students from study abroad participation, thus creating a climate where students do not view study abroad as an option that is possible.

Critical Values of the *t* Distribution

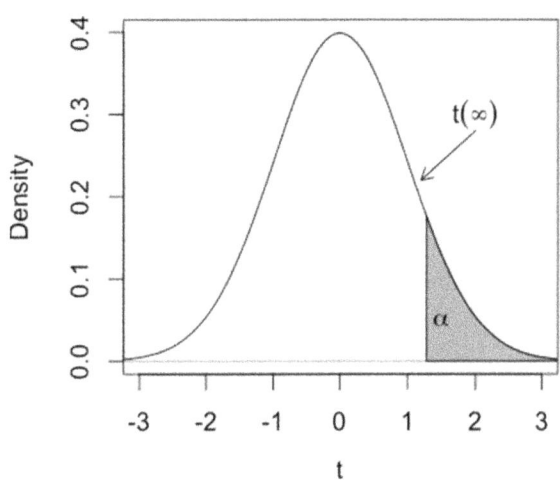

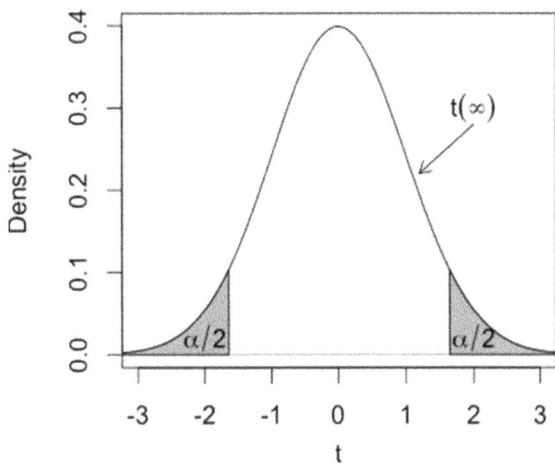

Appendix A: Critical Values of the t Distribution

	α level					
One-tail	0.1	0.05	0.025	0.01	0.005	0.001
Two-tail	0.2	0.10	0.050	0.02	0.010	0.002
df = 1	3.0777	6.3138	12.7062	31.8205	63.6567	318.3088
2	1.8856	2.9200	4.3027	6.9646	9.9248	22.3271
3	1.6377	2.3534	3.1824	4.5407	5.8409	10.2145
4	1.5332	2.1318	2.7764	3.7469	4.6041	7.1732
5	1.4759	2.0150	2.5706	3.3649	4.0321	5.8934
6	1.4398	1.9432	2.4469	3.1427	3.7074	5.2076
7	1.4149	1.8946	2.3646	2.9980	3.4995	4.7853
8	1.3968	1.8595	2.3060	2.8965	3.3554	4.5008
9	1.3830	1.8331	2.2622	2.8214	3.2498	4.2968
10	1.3722	1.8125	2.2281	2.7638	3.1693	4.1437
11	1.3634	1.7959	2.2010	2.7181	3.1058	4.0247
12	1.3562	1.7823	2.1788	2.6810	3.0545	3.9296
13	1.3502	1.7709	2.1604	2.6503	3.0123	3.8520
14	1.3450	1.7613	2.1448	2.6245	2.9768	3.7874
15	1.3406	1.7531	2.1314	2.6025	2.9467	3.7328
16	1.3368	1.7459	2.1199	2.5835	2.9208	3.6862
17	1.3334	1.7396	2.1098	2.5669	2.8982	3.6458
18	1.3304	1.7341	2.1009	2.5524	2.8784	3.6105
19	1.3277	1.7291	2.0930	2.5395	2.8609	3.5794
20	1.3253	1.7247	2.0860	2.5280	2.8453	3.5518
21	1.3232	1.7207	2.0796	2.5176	2.8314	3.5272
22	1.3212	1.7171	2.0739	2.5083	2.8188	3.5050
23	1.3195	1.7139	2.0687	2.4999	2.8073	3.4850
24	1.3178	1.7109	2.0639	2.4922	2.7969	3.4668
25	1.3163	1.7081	2.0595	2.4851	2.7874	3.4502
26	1.3150	1.7056	2.0555	2.4786	2.7787	3.4350
27	1.3137	1.7033	2.0518	2.4727	2.7707	3.4210
28	1.3125	1.7011	2.0484	2.4671	2.7633	3.4082
29	1.3114	1.6991	2.0452	2.4620	2.7564	3.3962
30	1.3104	1.6973	2.0423	2.4573	2.7500	3.3852
40	1.3031	1.6839	2.0211	2.4233	2.7045	3.3069
60	1.2958	1.6706	2.0003	2.3901	2.6603	3.2317
120	1.2886	1.6577	1.9799	2.3578	2.6174	3.1595
∞	**1.2816**	**1.6449**	**1.9600**	**2.3263**	**2.5758**	**3.0902**

Notes Cell entries are critical values for t(df) at the column's listed significance level. The last line of this table, with infinite degrees of freedom, represents the **Normal** distribution. Table produced in R 4.0.0

Critical Values of the F Distribution at α = 0.05

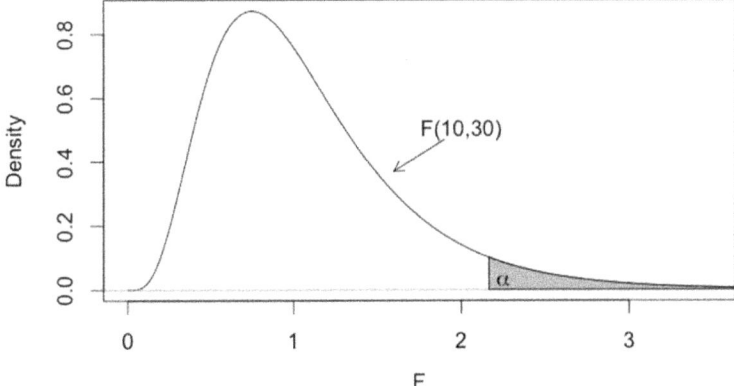

Note Subsequent cell entries are critical values for F(df1,df2) at the α = 0.05 significance level. Table produced in R 4.0.0

Appendix B: Critical Values of the F Distribution at $\alpha = 0.05$

Denominator DF (df2)	Numerator Degrees of Freedom (df1)									
	1	2	3	4	5	6	7	8	9	10
1	161.4476	199.5000	215.7073	224.5832	230.1619	233.9860	236.7684	238.8827	240.5433	241.8817
2	18.5128	19.0000	19.1643	19.2468	19.2964	19.3295	19.3532	19.3710	19.3848	19.3959
3	10.1280	9.5521	9.2766	9.1172	9.0135	8.9406	8.8867	8.8452	8.8123	8.7855
4	7.7086	6.9443	6.5914	6.3882	6.2561	6.1631	6.0942	6.0410	5.9988	5.9644
5	6.6079	5.7861	5.4095	5.1922	5.0503	4.9503	4.8759	4.8183	4.7725	4.7351
6	5.9874	5.1433	4.7571	4.5337	4.3874	4.2839	4.2067	4.1468	4.0990	4.0600
7	5.5914	4.7374	4.3468	4.1203	3.9715	3.8660	3.7870	3.7257	3.6767	3.6365
8	5.3177	4.4590	4.0662	3.8379	3.6875	3.5806	3.5005	3.4381	3.3881	3.3472
9	5.1174	4.2565	3.8625	3.6331	3.4817	3.3738	3.2927	3.2296	3.1789	3.1373
10	4.9646	4.1028	3.7083	3.4780	3.3258	3.2172	3.1355	3.0717	3.0204	2.9782
11	4.8443	3.9823	3.5874	3.3567	3.2039	3.0946	3.0123	2.9480	2.8962	2.8536
12	4.7472	3.8853	3.4903	3.2592	3.1059	2.9961	2.9134	2.8486	2.7964	2.7534
13	4.6672	3.8056	3.4105	3.1791	3.0254	2.9153	2.8321	2.7669	2.7144	2.6710
14	4.6001	3.7389	3.3439	3.1122	2.9582	2.8477	2.7642	2.6987	2.6458	2.6022
15	4.5431	3.6823	3.2874	3.0556	2.9013	2.7905	2.7066	2.6408	2.5876	2.5437
16	4.4940	3.6337	3.2389	3.0069	2.8524	2.7413	2.6572	2.5911	2.5377	2.4935
17	4.4513	3.5915	3.1968	2.9647	2.8100	2.6987	2.6143	2.5480	2.4943	2.4499
18	4.4139	3.5546	3.1599	2.9277	2.7729	2.6613	2.5767	2.5102	2.4563	2.4117

Appendix B: Critical Values of the F Distribution at $\alpha = 0.05$

Denominator DF (df2)	Numerator Degrees of Freedom (df1)									
	1	2	3	4	5	6	7	8	9	10
19	4.3807	3.5219	3.1274	2.8951	2.7401	2.6283	2.5435	2.4768	2.4227	2.3779
20	4.3512	3.4928	3.0984	2.8661	2.7109	2.5990	2.5140	2.4471	2.3928	2.3479
21	4.3248	3.4668	3.0725	2.8401	2.6848	2.5727	2.4876	2.4205	2.3660	2.3210
22	4.3009	3.4434	3.0491	2.8167	2.6613	2.5491	2.4638	2.3965	2.3419	2.2967
23	4.2793	3.4221	3.0280	2.7955	2.6400	2.5277	2.4422	2.3748	2.3201	2.2747
24	4.2597	3.4028	3.0088	2.7763	2.6207	2.5082	2.4226	2.3551	2.3002	2.2547
25	4.2417	3.3852	2.9912	2.7587	2.6030	2.4904	2.4047	2.3371	2.2821	2.2365
26	4.2252	3.3690	2.9752	2.7426	2.5868	2.4741	2.3883	2.3205	2.2655	2.2197
27	4.2100	3.3541	2.9604	2.7278	2.5719	2.4591	2.3732	2.3053	2.2501	2.2043
28	4.1960	3.3404	2.9467	2.7141	2.5581	2.4453	2.3593	2.2913	2.2360	2.1900
29	4.1830	3.3277	2.9340	2.7014	2.5454	2.4324	2.3463	2.2783	2.2229	2.1768
30	4.1709	3.3158	2.9223	2.6896	2.5336	2.4205	2.3343	2.2662	2.2107	2.1646
32	4.1491	3.2945	2.9011	2.6684	2.5123	2.3991	2.3127	2.2444	2.1888	2.1425
34	4.1300	3.2759	2.8826	2.6499	2.4936	2.3803	2.2938	2.2253	2.1696	2.1231
36	4.1132	3.2594	2.8663	2.6335	2.4772	2.3638	2.2771	2.2085	2.1526	2.1061
38	4.0982	3.2448	2.8517	2.6190	2.4625	2.3490	2.2623	2.1936	2.1375	2.0909
40	4.0847	3.2317	2.8387	2.6060	2.4495	2.3359	2.2490	2.1802	2.1240	2.0772
50	4.0343	3.1826	2.7900	2.5572	2.4004	2.2864	2.1992	2.1299	2.0734	2.0261
60	4.0012	3.1504	2.7581	2.5252	2.3683	2.2541	2.1665	2.0970	2.0401	1.9926
70	3.9778	3.1277	2.7355	2.5027	2.3456	2.2312	2.1435	2.0737	2.0166	1.9689
80	3.9604	3.1108	2.7188	2.4859	2.3287	2.2142	2.1263	2.0564	1.9991	1.9512

Appendix B: Critical Values of the F Distribution at α = 0.05

Denominator DF (df2)	Numerator Degrees of Freedom (df1)									
	1	2	3	4	5	6	7	8	9	10
100	3.9361	3.0873	2.6955	2.4626	2.3053	2.1906	2.1025	2.0323	1.9748	1.9267
125	3.9169	3.0687	2.6771	2.4442	2.2868	2.1719	2.0836	2.0133	1.9556	1.9072
150	3.9042	3.0564	2.6649	2.4320	2.2745	2.1595	2.0711	2.0006	1.9428	1.8943
200	3.8884	3.0411	2.6498	2.4168	2.2592	2.1441	2.0556	1.9849	1.9269	1.8783
400	3.8648	3.0183	2.6272	2.3942	2.2366	2.1212	2.0325	1.9616	1.9033	1.8544
1000	3.8508	3.0047	2.6138	2.3808	2.2231	2.1076	2.0187	1.9476	1.8892	1.8402
∞	3.8415	2.9957	2.6049	2.3719	2.2141	2.0986	2.0096	1.9384	1.8799	1.8307

Appendix B: Critical Values of the F Distribution at $\alpha = 0.05$

Denominator DF (df2)	Numerator Degrees of Freedom (df1)									
	11	12	13	14	15	16	17	18	19	20
1	242.9835	243.9060	244.6898	245.3640	245.9499	246.4639	246.9184	247.3232	247.6861	248.0131
2	19.4050	19.4125	19.4189	19.4244	19.4291	19.4333	19.4370	19.4402	19.4431	19.4458
3	8.7633	8.7446	8.7287	8.7149	8.7029	8.6923	8.6829	8.6745	8.6670	8.6602
4	5.9358	5.9117	5.8911	5.8733	5.8578	5.8441	5.8320	5.8211	5.8114	5.8025
5	4.7040	4.6777	4.6552	4.6358	4.6188	4.6038	4.5904	4.5785	4.5678	4.5581
6	4.0274	3.9999	3.9764	3.9559	3.9381	3.9223	3.9083	3.8957	3.8844	3.8742
7	3.6030	3.5747	3.5503	3.5292	3.5107	3.4944	3.4799	3.4669	3.4551	3.4445
8	3.3130	3.2839	3.2590	3.2374	3.2184	3.2016	3.1867	3.1733	3.1613	3.1503
9	3.1025	3.0729	3.0475	3.0255	3.0061	2.9890	2.9737	2.9600	2.9477	2.9365
10	2.9430	2.9130	2.8872	2.8647	2.8450	2.8276	2.8120	2.7980	2.7854	2.7740
11	2.8179	2.7876	2.7614	2.7386	2.7186	2.7009	2.6851	2.6709	2.6581	2.6464
12	2.7173	2.6866	2.6602	2.6371	2.6169	2.5989	2.5828	2.5684	2.5554	2.5436
13	2.6347	2.6037	2.5769	2.5536	2.5331	2.5149	2.4987	2.4841	2.4709	2.4589
14	2.5655	2.5342	2.5073	2.4837	2.4630	2.4446	2.4282	2.4134	2.4000	2.3879
15	2.5068	2.4753	2.4481	2.4244	2.4034	2.3849	2.3683	2.3533	2.3398	2.3275
16	2.4564	2.4247	2.3973	2.3733	2.3522	2.3335	2.3167	2.3016	2.2880	2.2756
17	2.4126	2.3807	2.3531	2.3290	2.3077	2.2888	2.2719	2.2567	2.2429	2.2304
18	2.3742	2.3421	2.3143	2.2900	2.2686	2.2496	2.2325	2.2172	2.2033	2.1906

Appendix B: Critical Values of the F Distribution at $\alpha = 0.05$

Denominator DF (df2)	Numerator Degrees of Freedom (df1)									
	11	12	13	14	15	16	17	18	19	20
19	2.3402	2.3080	2.2800	2.2556	2.2341	2.2149	2.1977	2.1823	2.1683	2.1555
20	2.3100	2.2776	2.2495	2.2250	2.2033	2.1840	2.1667	2.1511	2.1370	2.1242
21	2.2829	2.2504	2.2222	2.1975	2.1757	2.1563	2.1389	2.1232	2.1090	2.0960
22	2.2585	2.2258	2.1975	2.1727	2.1508	2.1313	2.1138	2.0980	2.0837	2.0707
23	2.2364	2.2036	2.1752	2.1502	2.1282	2.1086	2.0910	2.0751	2.0608	2.0476
24	2.2163	2.1834	2.1548	2.1298	2.1077	2.0880	2.0703	2.0543	2.0399	2.0267
25	2.1979	2.1649	2.1362	2.1111	2.0889	2.0691	2.0513	2.0353	2.0207	2.0075
26	2.1811	2.1479	2.1192	2.0939	2.0716	2.0518	2.0339	2.0178	2.0032	1.9898
27	2.1655	2.1323	2.1035	2.0781	2.0558	2.0358	2.0179	2.0017	1.9870	1.9736
28	2.1512	2.1179	2.0889	2.0635	2.0411	2.0210	2.0030	1.9868	1.9720	1.9586
29	2.1379	2.1045	2.0755	2.0500	2.0275	2.0073	1.9893	1.9730	1.9581	1.9446
30	2.1256	2.0921	2.0630	2.0374	2.0148	1.9946	1.9765	1.9601	1.9452	1.9317
32	2.1033	2.0697	2.0404	2.0147	1.9920	1.9717	1.9534	1.9369	1.9219	1.9083
34	2.0838	2.0500	2.0207	1.9949	1.9720	1.9516	1.9332	1.9166	1.9015	1.8877
36	2.0666	2.0327	2.0032	1.9773	1.9543	1.9338	1.9153	1.8986	1.8834	1.8696
38	2.0513	2.0173	1.9877	1.9616	1.9386	1.9179	1.8994	1.8826	1.8673	1.8534
40	2.0376	2.0035	1.9738	1.9476	1.9245	1.9037	1.8851	1.8682	1.8529	1.8389
50	1.9861	1.9515	1.9214	1.8949	1.8714	1.8503	1.8313	1.8141	1.7985	1.7841
60	1.9522	1.9174	1.8870	1.8602	1.8364	1.8151	1.7959	1.7784	1.7625	1.7480
70	1.9283	1.8932	1.8627	1.8357	1.8117	1.7902	1.7708	1.7531	1.7371	1.7223
80	1.9105	1.8753	1.8445	1.8174	1.7932	1.7716	1.7520	1.7342	1.7180	1.7032

Appendix B: Critical Values of the F Distribution at α = 0.05

Denominator DF (df2)	Numerator Degrees of Freedom (df1)									
	11	12	13	14	15	16	17	18	19	20
100	1.8857	1.8503	1.8193	1.7919	1.7675	1.7456	1.7259	1.7079	1.6915	1.6764
125	1.8660	1.8304	1.7992	1.7717	1.7471	1.7250	1.7051	1.6869	1.6704	1.6551
150	1.8530	1.8172	1.7859	1.7582	1.7335	1.7113	1.6913	1.6730	1.6563	1.6410
200	1.8368	1.8008	1.7694	1.7415	1.7166	1.6943	1.6741	1.6556	1.6388	1.6233
400	1.8126	1.7764	1.7447	1.7166	1.6914	1.6688	1.6484	1.6297	1.6126	1.5969
1000	1.7982	1.7618	1.7299	1.7017	1.6764	1.6536	1.6330	1.6142	1.5969	1.5811
∞	1.7886	1.7522	1.7202	1.6918	1.6664	1.6435	1.6228	1.6038	1.5865	1.5705

Appendix B: Critical Values of the F Distribution at α = 0.05

Denominator DF (df2)	Numerator Degrees of Freedom (df1)										
	22	24	26	28	30	40	50	100	200	∞
1	248.5791	249.0518	249.4525	249.7966	250.0951	251.1432	251.7742	253.0411	253.6770	254.3144
2	19.4503	19.4541	19.4573	19.4600	19.4624	19.4707	19.4757	19.4857	19.4907	19.4957
3	8.6484	8.6385	8.6301	8.6229	8.6166	8.5944	8.5810	8.5539	8.5402	8.5264
4	5.7872	5.7744	5.7635	5.7541	5.7459	5.7170	5.6995	5.6641	5.6461	5.6281
5	4.5413	4.5272	4.5151	4.5047	4.4957	4.4638	4.4444	4.4051	4.3851	4.3650
6	3.8564	3.8415	3.8287	3.8177	3.8082	3.7743	3.7537	3.7117	3.6904	3.6689
7	3.4260	3.4105	3.3972	3.3858	3.3758	3.3404	3.3189	3.2749	3.2525	3.2298
8	3.1313	3.1152	3.1015	3.0897	3.0794	3.0428	3.0204	2.9747	2.9513	2.9276
9	2.9169	2.9005	2.8864	2.8743	2.8637	2.8259	2.8028	2.7556	2.7313	2.7067
10	2.7541	2.7372	2.7229	2.7104	2.6996	2.6609	2.6371	2.5884	2.5634	2.5379
11	2.6261	2.6090	2.5943	2.5816	2.5705	2.5309	2.5066	2.4566	2.4308	2.4045
12	2.5229	2.5055	2.4905	2.4776	2.4663	2.4259	2.4010	2.3498	2.3233	2.2962
13	2.4379	2.4202	2.4050	2.3918	2.3803	2.3392	2.3138	2.2614	2.2343	2.2064
14	2.3667	2.3487	2.3333	2.3199	2.3082	2.2664	2.2405	2.1870	2.1592	2.1307
15	2.3060	2.2878	2.2722	2.2587	2.2468	2.2043	2.1780	2.1234	2.0950	2.0658
16	2.2538	2.2354	2.2196	2.2059	2.1938	2.1507	2.1240	2.0685	2.0395	2.0096
17	2.2084	2.1898	2.1738	2.1599	2.1477	2.1040	2.0769	2.0204	1.9909	1.9604
18	2.1685	2.1497	2.1335	2.1195	2.1071	2.0629	2.0354	1.9780	1.9479	1.9168

Appendix B: Critical Values of the F Distribution at $\alpha = 0.05$

Denominator DF (df2)	Numerator Degrees of Freedom (df1)										
	22	24	26	28	30	40	50	100	200	∞
19	2.1331	2.1141	2.0978	2.0836	2.0712	2.0264	1.9986	1.9403	1.9097	1.8780
20	2.1016	2.0825	2.0660	2.0517	2.0391	1.9938	1.9656	1.9066	1.8755	1.8432
21	2.0733	2.0540	2.0374	2.0229	2.0102	1.9645	1.9360	1.8761	1.8446	1.8117
22	2.0478	2.0283	2.0116	1.9970	1.9842	1.9380	1.9092	1.8486	1.8165	1.7831
23	2.0246	2.0050	1.9881	1.9734	1.9605	1.9139	1.8848	1.8234	1.7909	1.7570
24	2.0035	1.9838	1.9668	1.9520	1.9390	1.8920	1.8625	1.8005	1.7675	1.7330
25	1.9842	1.9643	1.9472	1.9323	1.9192	1.8718	1.8421	1.7794	1.7460	1.7110
26	1.9664	1.9464	1.9292	1.9142	1.9010	1.8533	1.8233	1.7599	1.7261	1.6906
27	1.9500	1.9299	1.9126	1.8975	1.8842	1.8361	1.8059	1.7419	1.7077	1.6717
28	1.9349	1.9147	1.8973	1.8821	1.8687	1.8203	1.7898	1.7251	1.6905	1.6541
29	1.9208	1.9005	1.8830	1.8677	1.8543	1.8055	1.7748	1.7096	1.6746	1.6376
30	1.9077	1.8874	1.8698	1.8544	1.8409	1.7918	1.7609	1.6950	1.6597	1.6223
32	1.8842	1.8636	1.8458	1.8303	1.8166	1.7670	1.7356	1.6687	1.6326	1.5943
34	1.8634	1.8427	1.8248	1.8091	1.7953	1.7451	1.7134	1.6454	1.6086	1.5694
36	1.8451	1.8242	1.8061	1.7904	1.7764	1.7257	1.6936	1.6246	1.5872	1.5471
38	1.8288	1.8077	1.7895	1.7736	1.7596	1.7084	1.6759	1.6060	1.5679	1.5271
40	1.8141	1.7929	1.7746	1.7586	1.7444	1.6928	1.6600	1.5892	1.5505	1.5089
50	1.7588	1.7371	1.7183	1.7017	1.6872	1.6337	1.5995	1.5249	1.4835	1.4383
60	1.7222	1.7001	1.6809	1.6641	1.6491	1.5943	1.5590	1.4814	1.4377	1.3893
70	1.6962	1.6738	1.6543	1.6372	1.6220	1.5661	1.5300	1.4498	1.4042	1.3529
80	1.6768	1.6542	1.6345	1.6171	1.6017	1.5449	1.5081	1.4259	1.3786	1.3247

Appendix B: Critical Values of the F Distribution at $\alpha = 0.05$

Denominator DF (df2)	Numerator Degrees of Freedom (df1)										
	22	24	26	28	30	40	50	100	200	∞
100	1.6497	1.6267	1.6067	1.5890	1.5733	1.5151	1.4772	1.3917	1.3416	1.2832
125	1.6281	1.6048	1.5844	1.5665	1.5505	1.4912	1.4524	1.3638	1.3110	1.2478
150	1.6137	1.5902	1.5696	1.5515	1.5354	1.4752	1.4357	1.3448	1.2899	1.2226
200	1.5958	1.5720	1.5511	1.5328	1.5164	1.4551	1.4146	1.3206	1.2626	1.1885
400	1.5689	1.5446	1.5234	1.5046	1.4878	1.4247	1.3827	1.2831	1.2189	1.1279
1000	1.5528	1.5282	1.5067	1.4876	1.4706	1.4063	1.3632	1.2596	1.1903	1.0781
∞	1.5420	1.5173	1.4956	1.4763	1.4591	1.3940	1.3501	1.2434	1.1700	1.0000

Critical Values of the Chi-Square Distribution

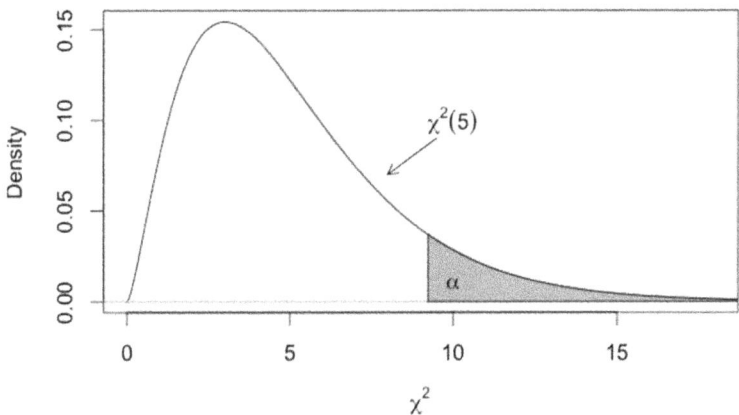

df	α level			
	0.1	0.05	0.01	0.001
1	2.7055	3.8415	6.6349	10.8276
2	4.6052	5.9915	9.2103	13.8155
3	6.2514	7.8147	11.3449	16.2662
4	7.7794	9.4877	13.2767	18.4668
5	9.2364	11.0705	15.0863	20.5150
6	10.6446	12.5916	16.8119	22.4577
7	12.0170	14.0671	18.4753	24.3219
8	13.3616	15.5073	20.0902	26.1245
9	14.6837	16.9190	21.6660	27.8772
10	15.9872	18.3070	23.2093	29.5883

© The Editor(s) (if applicable) and The Author(s), under exclusive license to Springer Nature Switzerland AG 2022
M. Whatley, *Introduction to Quantitative Analysis for International Educators*, Springer Texts in Education, https://doi.org/10.1007/978-3-030-93831-4

Appendix C: Critical Values of the Chi-Square Distribution

	α level			
df	0.1	0.05	0.01	0.001
11	17.2750	19.6751	24.7250	31.2641
12	18.5493	21.0261	26.2170	32.9095
13	19.8119	22.3620	27.6882	34.5282
14	21.0641	23.6848	29.1412	36.1233
15	22.3071	24.9958	30.5779	37.6973
16	23.5418	26.2962	31.9999	39.2524
17	24.7690	27.5871	33.4087	40.7902
18	25.9894	28.8693	34.8053	42.3124
19	27.2036	30.1435	36.1909	43.8202
20	28.4120	31.4104	37.5662	45.3147
21	29.6151	32.6706	38.9322	46.7970
22	30.8133	33.9244	40.2894	48.2679
23	32.0069	35.1725	41.6384	49.7282
24	33.1962	36.4150	42.9798	51.1786
25	34.3816	37.6525	44.3141	52.6197
26	35.5632	38.8851	45.6417	54.0520
27	36.7412	40.1133	46.9629	55.4760
28	37.9159	41.3371	48.2782	56.8923
29	39.0875	42.5570	49.5879	58.3012
30	40.2560	43.7730	50.8922	59.7031

Notes Cell entries are critical values for χ^2 (df) at the column's listed significance level. Table produced in R 4.0.0

Current International Education (and Related) Journals

Compare: A Journal of Comparative and International Education
Comparative and International Education/Éducation Comparée et Internationale
Comparative Education
Comparative Education Review
Frontiers: The Interdisciplinary Journal of Study Abroad
Globalisation, Societies, and Education
Higher Education
International Education Journal: Comparative Perspectives
International Journal of Educational Development
International Journal of Intercultural Relations
Journal for the Study of Postsecondary and Tertiary Education
Journal of Comparative and International Higher Education
Journal of Higher Education Policy and Management
Journal of International Students
Journal of Research in International Education
Journal of Studies in International Education
Minerva: A Review of Science, Learning, and Policy
Studies in Higher Education

Glossary

Alpha level A value that the researcher selects for the purpose of hypothesis testing that indicates the percentage of the time a Type 1 error is acceptable. Common alpha levels in education research include 0.05, 0.01, and 0.001.

Alternative hypothesis The hypothesis that there is a statistically significant relationship in a comparison of interest

Asymptotic A property of the normal distribution indicating that its tails never quite touch the horizontal axis, although they get very close

Attrition A threat to a randomized control trial that happens when individuals in the treatment and control group drop out of the study in a way that is not random

Average marginal effect The average of discrete changes in the predicted probability of a given outcome; one of the possible transformations for logistic regression coefficients

Categorical variable A variable that has categories (e.g., demographic information)

Causal inference A type of research design that allows the researcher to isolate, at least theoretically, the causal link between a treatment condition and a specific outcome

Chi-square test of independence A hypothesis test that compares two categorical variables

Coefficient of determination The percentage of variance that is shared between two variables that are correlated

Common support In propensity score matching, the area where treatment and control group propensity scores overlap

Comparison category See Reference category

Confidence interval A range of values that likely contains a population mean, calculated using a sample mean, estimated standard error, and critical t-value that is associated with the alpha level that the researcher chooses

Contamination A threat to a randomized control trial that happens when individuals in the treatment and control groups are able to interact with one another

Continuous variable A numerical variable that indicates a specific quantity (e.g., a country's GDP)

Convenience sampling Sampling in a way that collects data from participants that are in close proximity to the researcher who are also willing to participate

Correlation The standardized joint variability of two variables

Correlation coefficient A numerical summary of the strength of the relationship between two continuous variables

Covariance The unstandardized joint variability of two variables

Covariate See Explanatory variable

Critical value A value in a particular distribution (e.g., t-distribution, F-distribution, chi-square distribution) that is used as a reference point in hypothesis testing

Dependent samples t-test A t-test that tests the difference in sample means for the same units over time, thus comparing sample means at two time points

Dependent variable See Outcome variable

Descriptive statistics Statistics that are used to describe a particular sample

Dichotomous variable A categorical variable that has only two categories (e.g., if a country's official language is English—yes or no)

Difference-in-differences analysis A type of quasi-experimental design that relies on a policy or practice that takes place at a specific point in time and that sorts individuals into treatment and control groups

Distribution (of a variable) Frequency with which each value of a variable appears in the dataset

Dummy variable Variables in a regression model a provide categorical information about units in a dataset (e.g., a student's racial or ethnic identity)

Error In OLS regression, the difference between predicted and observed values

Expected value of the mean The mean of a sampling distribution and the best approximation of the population mean

Explanatory variable A variable that a researcher uses to explain or predict the outcome variable in a regression analysis

Fidelity of implementation A threat to a randomized control trial that happens when a given intervention is not implemented as intended

Field data See Secondary data

Functional form The shape of the line that depicts the relationship between two variables in a regression model

Grand mean The overall mean, which includes data from all groups, in an ANOVA analysis

Histogram A common way to visualize a variable's distribution with the vertical axis representing the frequency with which certain values appear in a dataset and the horizontal axis representing counts of units, divided into categories or bins, that are of equal sizes

Independent variable See Explanatory variable

Inferential statistics Statistics based on a sample that can generalize to a population

Interaction variable A variable in a regression model that is used to explore whether the relationship between a predictor variable and the outcome variable changes depending on the value of another predictor variable.

Intercept The point where a regression line crosses the y-axis (the predicted value of the outcome variable if all predictor variables are set to zero)

Interval/ratio variable A numerical variable that can be ordered meaningfully and that exhibits distances between values that are uniform (e.g., measures of distance)

Left skewed A distribution with only a few values that fall along the left-hand side of the distribution, meaning that there are only a few low values

Linear probability model A linear regression model with a binary outcome variable

Log odds The default unit for coefficients in a logistic regression model; the log transformation of the odds that an event or situation occurs

Logistic regression An approach to regression modeling, estimated using maximum likelihood, that is appropriate for binary outcome variables

Mean The arithmetic average of a variable's distribution

Mean square between The average amount of variation between groups (used in an ANOVA analysis)

Mean square within The average amount of variation within groups (used in an ANOVA analysis)

Median The number in the middle of a variable's distribution

Mode The most frequent number in a variable's distribution

Multicollinearity A problem in multiple regression analysis where two independent variables are highly correlated with one another

Multiple linear regression A type of regression analysis that uses more than one variable to predict a continuous outcome variable

Negatively skewed See Left skewed

Nominal variable A categorical variable wherein numbers are representative of categories but are otherwise not meaningful (e.g., numbers that represent a student's home region of the world)

Non-linear functional form A functional form depicting the relationship between two variables that is not a straight line

Nonparametric A family of statistics that does not rely on parameters (e.g., mean, variance) and therefore allows for the relaxing of assumptions of parametric statistics (e.g., that a sampling distribution approximates a normal distribution)

Normal distribution A hypothetical distribution that is symmetrical, unimodal, and asymptotic and that incorporates under its curve a certain percentage of observations within one, two, and three standard deviations

Null hypothesis The hypothesis that there is no statistically significant relationship between two means

Observational data See Secondary data

Observed value An actual value of a variable found in a dataset (as opposed to a predicted value)

Observed variable A variable in a dataset that has been collected and whose value is known to the researcher

Odds ratio The ratio of the probability that an event will occur to the probability that an event will not occur; one of the possible transformations for logistic regression coefficients

One-sample t-test A t-test that compares a sample mean with a hypothesized population mean

One-tailed hypothesis test A hypothesis test that accounts for the area in only one tail of the normal or t-distribution

One-way analysis of variance (ANOVA) A statistical test that compares sample means of two or more groups

Ordinal variable A numerical variable that indicates categories that are meaningful (e.g., Likert-scale survey responses)

Ordinary least squares An approach to regression that involves finding the regression line that minimizes the sum of squared errors between observed values of an outcome variable and the regression line

Outcome variable The variable that a researcher wants to explain in a regression analysis

Outlier A value in a variable's distribution that is much lower or much higher than most of the other values in that distribution

P-value The probability of obtaining a particular sample mean if the null hypothesis is true

Parameter A quantifiable characteristic of a population (e.g., mean, variance)

Parametric A family of statistics that relies on parameters (e.g., mean, variance) to construct sampling distributions and test statistics

Pearson product-moment correlation coefficient A correlation coefficient that summarizes the strength of the relationship between two interval or ratio variables

Population The complete group of units (individuals, institutions, countries, etc.) that is the focus of a particular study

Positively skewed See Right skewed

Predicted value In regression, the value of the dependent variable that is predicted when independent variables are set to certain values

Predictor variable See Explanatory variable

Propensity score A measure belonging to a unit in a dataset that represents the unit's likelihood of belonging to the treatment group based on the dataset's pretreatment variables

Propensity score matching An analytic approach that relies on matching treatment and control students with one another in a way that reduces the differences between the two groups along observed characteristics

Quadratic functional form A non-linear functional form that is u- or inverse-u-shaped; in practice, a quadratic functional form involves squaring the predictor of interest in a regression model

Qualitative variable See Dummy variable

Quasi-experimental design A type of causal inference that makes use of statistical approaches to correct for the lack of randomization of treatment and control conditions

R-squared A measure of model fit in ordinary least squares regression that provides the percentage of variance in the dependent variable that is explained by the model's independent variables

Random sampling Sampling in a way that each member of a study's population of interest has equal likelihood of being selected to participate in the study

Randomized control trial A type of causal inference study that makes use of randomization to isolate the treatment effect

Range The maximum value of a variable minus its minimum value

Ratio/interval variable See Interval/ratio variable

Reference category In a regression model, the category coded '0' (e.g., groups of racial/ethnic identities), which is used to interpret results

Regression coefficient See Slope

Regression discontinuity A type of quasi-experimental design that relies on a running variable, such as a test score, that is used to place individuals into treatment and control groups

Regression line A line drawn through the data points on a scatterplot that represents the average relationship between two variables

Representative sampling Sampling in a way that purposefully selects certain individuals or units from the population so that the larger population is proportionally represented

Residual See Error

Right skewed A distribution with only a few values that fall along the right-hand side of the distribution, meaning that there are only a few high values

Sample The subset of a population that is used in a specific research study

Sampling distribution of the mean The distribution of means of a particular variable that results from taking multiple random samples from the same population. According to the Central Limit Theorem, this distribution approximates a normal distribution.

Sampling strategy The process that a researcher uses for collecting data from a subset of the population of interest in a given study

Scatterplot A visual depiction of the relationship between two variables where each individual observation is represented with a dot placed at the intersection of the x- and y-axes corresponding to its particular values for both variables

Secondary data Data that were collected for other purposes, but that can be used to inform additional research questions

Selection bias Bias that occurs in an analysis for causal inference due to significant differences between treatment and control groups along the lines of unobserved characteristics that are very likely related to the outcome of interest

Simple linear regression A type of regression analysis that uses one variable to predict a continuous outcome variable

Slope In reference to a regression line, the ratio of the vertical change to the horizontal change in the line, which provides the predicted average increase in the dependent variable if the independent variable increases by one unit

Standard deviation A measure of the dispersion of values around the mean of a distribution in the original scale of the variable

Standard error A standardized measure of variability in a sampling distribution that provides information about the variability among sample means

Sum of squares between groups The total amount of variation between groups (used in an ANOVA analysis)

Sum of squares within groups The total amount of variation within groups (used in an ANOVA analysis)

Symmetrical A property of the normal distribution indicating that data are distributed so that the frequency of values on either side of the mean are a reflection of one another

Two-samples t-test A t-test that examines the difference in two sample means

Two-tailed hypothesis test A hypothesis test that accounts for the area in both tails of the normal or t-distribution

Type 1 error Detection of a significant difference in means when there is no significant difference in reality

Type 2 error Failure to detect a significant difference in means when one exists in reality

Unimodal A property of the normal distribution indicating that it has only one most frequent value (mode), which is also the mean and the median

Unobserved variable A variable that is not in a dataset, has not been collected, and thus whose value is unknown to the researcher

Variable Information about a certain property or characteristic that corresponds to individual units (or observations) in a dataset

Variance A measure of the dispersion of values around the mean of a distribution in squared deviation units

Bibliography

Agresti, A., & Finlay, B. (2009). *Statistical methods for the social sciences*. Pearson.
Baruffaldi, S. H., Marino, M., & Visentin, F. (2020). Money to move: The effect on researchers of an international mobility grant. *Research Policy, 49*(8), 1–18.
Bochner, S., McLeod, B. M., & Lin, A. (1977). Friendship patterns of overseas students: A functional model. *International Journal of Psychology, 12*(4), 277–294.
Brunsting, N. C., Smith, A. C., & Zachry, C. E. (2018). An academic and cultural transition course for international students: Efficacy and socio-emotional outcomes. *Journal of International Students, 8*(4), 1497–1521.
Cartwright, C., Stevens, M., & Schneider, K. (2021). Constructing the learning outcomes with intercultural assessment: A 3-year study of a graduate study abroad and global experience programs. *Frontiers: The Interdisciplinary Journal of Study Abroad, 33*(1), 82–105.
Cox, B. E., McIntosh, K., Reason, R. D., & Terenzini, P. T. (2014). Working with missing data in higher education research: A primer and real-world example. *The Review of Higher Education, 37*(3), 377–402.
Cunningham, S. (2021). *Causal inference: The Mixtape*. Yale University Press.
d'Hombres, B., & Schnepf, S. V. (2021). International mobility of students in Italy and the UK: Does it pay off and for whom? *Higher Education*. https://doi.org/10.1007/s10734-020-00631-1
De Leeuw, E. D., Hox, J., & Dillman, D. (2012). *International handbook of survey methodology*. Routledge.
Di Pietro, G. (2020). Does an international academic environment promote study abroad? *Journal of Studies in International Education*, 1028315320913260.
Dicks, A., & Lancee, B. (2018). Double disadvantage in school? Children of immigrants and the relative age effect: A regression discontinuity design based on the month of birth. *European Sociological Review, 34*(3), 319–333.
Echcharfy, M. (2020). Intercultural learning in Moroccan higher education: A comparison between teachers' perceptions and students' expectations. *International Journal of Research in English Education, 5*(1), 19–35.
Finn, M., Mihut, G., & Darmody, M. (2021). Academic satisfaction of international students at irish higher education institutions: The role of region of origin and cultural distance in the context of marketization. *Journal of Studies in International Education*, 10283153211027009.
Gujarati, D. N., & Porter, D. C. (2010). *Essentials of Econometrics* (4th ed.). McGraw-Hill.
Güzel, H., & Glazer, S. (2019). Demographic correlates of acculturation and sociocultural adaptation: Comparing international and domestic students. *Journal of International Students, 9*(4), 1074–1094.
Haupt, J. P. (2021). Short-term internationally mobile academics and their research collaborations upon return: Insights from the fulbright us scholar program. *Journal of Studies in International Education*, 1028315321990760.
Helms, R.M., Brajkovic, L., & Struthers, B. (2017). *Mapping internationalization on U.S. campuses: 2017 edition*. Retrieved from https://www.acenet.edu/Documents/Mapping-Internationalization-2017.pdf

Hendrickson, B., Rosen, D., & Aune, R. K. (2011). An analysis of friendship networks, social connectedness, homesickness, and satisfaction levels of international students. *International Journal of Intercultural Relations, 35*, 281–295.

Ibrahim, A. (2018). The happiness of undergraduate students at one university in the United Arab Emirates. *International Journal of Research Studies in Education, 7*(3), 49–61.

Iriondo, I. (2020). Evaluation of the impact of Erasmus study mobility on salaries and employment of recent graduates in Spain. *Studies in Higher Education, 45*(4), 925–943.

Jin, L., & Schneider, J. (2019). Faculty views on international students: A survey study. *Journal of International Students, 9*(1), 84–99.

Kritz, M. M. (2016). Why do countries differ in their rates of outbound student mobility? *Journal of Studies in International Education, 20*(2), 99–117.

Lee, D., Allen, M., Cheng, L., Watson, S., & Watson, W. (2021). Exploring relationships between self-efficacy and self-regulated learning strategies of english language learners in a college setting. *Journal of International Students, 11*(3), 567–585.

Lingo, M. D. (2019). Stratification in study abroad participation after accounting for student intent. *Research in Higher Education, 60*(8), 1142–1170.

Manly, C. A., & Wells, R. S. (2015). reporting the use of multiple imputation for missing data in higher education research. *Research in Higher Education, 56*, 397–409.

Marini, G., & Yang, L. (2021). Globally bred chinese talents returning home: An analysis of a reverse brain-drain flagship policy. *Science and Public Policy.* https://doi.org/10.1093/scipol/scab021

Markham, P., Rice, M., Darban, B., & Weng, T. H. (2017). Teachers' declared intentions to shift practice to incorporate second language acquisition (SLA) theories. *Journal of Language Teaching and Research, 8*(6), 1023–1031.

Martinsen, R. A. (2010). Short-term study abroad: Predicting changes in oral skills. *Foreign Language Annals, 43*(3), 504–530.

Meegan, C. K., & Kashima, E. S. (2010). Emotional and self-esteem consequences of perceiving discrimination against a new identity group. *Asian Journal of Social Psychology, 13*, 195–201.

Monogan, J. E., & Doctor, A. C. (2017). Immigration politics and partisan realignment: California, Texas, and the 1994 election. *State Politics & Policy Quarterly, 17*(1), 3–23.

OECD. (2019). PISA results (Volume 1): What students know and can do. *PISA, OECD Publishing.* https://doi.org/10.1787/5f07c754-en

Park, E. (2019). Issues of international students' academic adaptation in the ESL writing class: A mixed-methods study. *Journal of International Students, 2016 Vol. 6* (4), 887–904.

Schmidt, A. (2020). Are international students getting a bang for their buck? The relationship between expenditures and international student graduation rates. *Journal of International Students, 10*(3), 646–663.

Strange, H., & Gibson, H. J. (2017). An investigation of experiential and transformative learning in study abroad programs. *Frontiers: The Interdisciplinary Journal of Study Abroad, 29*(1), 85–100.

Urdan, T. C. (2017). *Statistics in plain English.* Routledge.

Ward, C., Bochner, S., & Furnham, A. (2001). *The psychology of culture shock* (2nd ed). Routledge.

Whatley, M. (2019). Study abroad participation: An unintended consequence of state merit-aid programs? *Research in Higher Education, 60*(7), 905–930.

Wheelan, C. (2013). *Naked statistics: Stripping the dread from data.* Norton.

Wooldridge, J. M. (2016). *Introductory econometrics: A modern approach.* Cengage Learning.

GPSR Compliance

The European Union's (EU) General Product Safety Regulation (GPSR) is a set of rules that requires consumer products to be safe and our obligations to ensure this.

If you have any concerns about our products, you can contact us on

ProductSafety@springernature.com

In case Publisher is established outside the EU, the EU authorized representative is:

Springer Nature Customer Service Center GmbH
Europaplatz 3
69115 Heidelberg, Germany

www.ingramcontent.com/pod-product-compliance
Ingram Content Group UK Ltd.
Pitfield, Milton Keynes, MK11 3LW, UK
UKHW022231230426
12048UKWH00016BA/1182